Recent Advances in Membrane Filtration and Purification Technologies

Recent Advances in Membrane Filtration and Purification Technologies

Guest Editor

Ahmed M. Khalil

MDPI

Basel • Beijing • Wuhan • Barcelona • Belgrade • Novi Sad • Cluj • Manchester

Guest Editor
Ahmed M. Khalil
Photochemistry Department
National Research Centre (NRC)
Giza
Egypt

Editorial Office
MDPI AG
Grosspeteranlage 5
4052 Basel, Switzerland

This is a reprint of the Special Issue, published open access by the journal *Membranes* (ISSN 2077-0375), freely accessible at: https://www.mdpi.com/journal/membranes/special_issues/Mem_Filtration_Purification.

For citation purposes, cite each article independently as indicated on the article page online and as indicated below:

Lastname, A.A.; Lastname, B.B. Article Title. *Journal Name* **Year**, *Volume Number*, Page Range.

ISBN 978-3-7258-7597-9 (Hbk)
ISBN 978-3-7258-7598-6 (PDF)
https://doi.org/10.3390/books978-3-7258-7598-6

Contents

About the Editor

Ahmed M. Khalil

Ahmed M. Khalil is a senior researcher at the Photochemistry Department - National Research Center (NRC), Egypt. He is dedicated to the fields of polymer science and photochemistry. His work focuses on enhancing the performance of polymers through advanced tailoring techniques, including the development of membranes for water remediation and composites modified with nanomaterials or antimicrobial substrates. These innovations aim to improve material functionality for a wide range of applications, reflecting his commitment to both scientific rigor and applied relevance. In addition to his research, he has made significant contributions to numerous peer-reviewed publications, patents, and collaborative projects focusing on polymer recycling and waste management for a greener environment. His efforts are driven by a vision to advance materials science to support ecological sustainability and industrial innovation, bridging fundamental research with practical solutions.

Preface

In recent decades, membrane innovation has flourished in water treatment, desalination, and industrial separation through filtration, reverse osmosis and gas separation. However, high processing costs limit widespread use, driving technology toward more efficient and affordable solutions. Recent advances in membrane design promise sustainable applications, yet fouling remains a major challenge. This reprint explores advancements in materials, system design and feasible applications. It offers fresh insights into separation, purification and environmental sustainability.

Ahmed M. Khalil
Guest Editor

membranes MDPI

Editorial

Recent Advances in Membrane Filtration and Purification Technologies

Ahmed M. Khalil

Photochemistry Department, National Research Centre, Dokki, Giza 12622, Egypt; akhalil75@yahoo.com

Membrane technologies comprise a broad range of materials and structural designs tailored for specific separation applications, including gas separation [1], water purification, wastewater treatment, and energy-related processes. Based on material composition, membranes can be broadly classified as polymeric, inorganic (ceramic and metallic) or hybrid membranes, each offering distinct advantages and limitations. Polymeric membranes dominate large-scale applications due to their low cost, ease of fabrication, and tunable physicochemical properties, making them widely used in water remediation processes such as microfiltration, ultrafiltration, nanofiltration, desalination and reverse osmosis. In contrast, inorganic membranes, comprising ceramic and metallic membranes, exhibit superior thermal, chemical, and mechanical stability. Hence, they are suitable for harsh operating environments. Hybrid systems, such as mixed-matrix membranes that combine polymers with inorganic fillers, have emerged as promising solutions to overcome the permeability and selectivity trade-off while expanding the applicability of membrane technologies across diverse industrial sectors [2,3]. In application-specific contexts, membrane design is closely aligned with separation mechanisms and operating requirements. For water remediation, polymeric and polymer-based composite membranes are predominantly employed to remove suspended solids, organic pollutants, dyes and heavy metals through size exclusion, charge-based repulsion and affinity-driven interactions. Surface modification and nanocomposite strategies are increasingly used to enhance fouling resistance and contaminant selectivity. Conversely, hydrogen separation membranes demand materials with exceptional hydrogen permeability and selectivity, leading to the historical dominance of dense metallic membranes, particularly palladium-based alloys, as well as the growing interest in ceramic proton-conducting membranes and advanced alloy systems. Recent research efforts are converging toward multifunctional membrane platforms that integrate high selectivity, durability, and cost effectiveness, enabling membranes to play a critical role in both clean energy production and sustainable water treatment, two of the most pressing technological challenges of the modern era [4].

Hydrogen separation membranes are critical components in clean energy systems, particularly for purifying hydrogen produced from steam reforming, coal gasification, methane pyrolysis, and water electrolysis processes. These membranes enable selective permeation of hydrogen over other gases (e.g., CO_2, N_2, CH_4), offering an energy-efficient alternative to conventional methods such as cryogenic distillation and pressure swing adsorption. While dense metallic-based alloys have historically dominated due to their high solubility and selectivity for hydrogen, they face challenges including high cost and susceptibility to poisoning [5]. Consequently, polymeric membranes, mixed-matrix membranes (MMMs) that integrate inorganic fillers into polymer matrices, and other advanced gas-separation materials are being developed to balance permeability and selectivity with lower fabrication costs and improved mechanical properties. These materials are engineered to enhance H_2 selectivity while maintaining robust performance under relevant operating conditions, positioning membrane technology as a key enabler in hydrogen purification for sustainable energy systems [6].

The increasing scarcity of clean water resources has emerged as one of the most critical global challenges of the twenty-first century, driven by rapid industrialization, urban expansion, and population growth. Industrial effluents containing organic pollutants, synthetic dyes, and toxic heavy metals are major contributors to water contamination, posing severe risks to ecosystems and human health. Conventional water treatment technologies, such as coagulation, adsorption, and biological treatment, often suffer from limitations related to efficiency, selectivity, and operational cost. In this context, membrane-based separation processes have gained significant attention as promising and sustainable alternatives for water remediation [7,8].

Composite membranes, in particular, have become principal to modern water treatment technologies due to their tunable physicochemical properties, scalability, and cost-effectiveness. Moreover, polymeric membranes function as selective barriers that allow water molecules to pass while rejecting contaminants based on size exclusion, charge interactions or affinity-based mechanisms. Compared with ceramic and metallic membranes, polymeric membranes offer advantages such as flexibility in fabrication, lower energy requirements, and ease of surface modification. As a result, they have been extensively investigated for applications in microfiltration, ultrafiltration, nanofiltration, and reverse osmosis processes [9,10].

Water pollutants encompass a wide range of contaminants, including suspended solids, organic compounds, dyes, pharmaceuticals, and heavy metals. Many of these substances are persistent, non-biodegradable, and capable of bioaccumulation, leading to long-term environmental consequences. Polymeric membranes provide an effective platform for addressing this complexity, as their separation performance can be tailored by controlling pore size, surface charge, hydrophilicity and membrane morphology. This adaptability makes them particularly suitable for treating industrial wastewater streams with variable compositions. Synthetic dyes represent a major class of water pollutants, especially in effluents from textile, leather, paper, and dyeing industries. These compounds are typically characterized by complex aromatic structures, high solubility, and resistance to degradation. Even at low concentrations, dyes impart intense color to water bodies, reducing light penetration and disrupting aquatic ecosystems. Polymeric membranes, especially nanofiltration and ultrafiltration membranes, have demonstrated high dye rejection efficiencies through size exclusion and electrostatic interactions, offering an effective means of decolorization and pollutant removal. In addition to dyes, heavy metals such as lead, cadmium, chromium, mercury, and arsenic are among the most hazardous water contaminants due to their toxicity, persistence, and tendency to accumulate in living organisms. Industrial activities including mining, electroplating, battery manufacturing, and metal finishing are primary sources of heavy metal pollution. Polymeric membranes can effectively remove heavy metal ions by mechanisms such as complexation with functional groups and adsorption onto modified membrane surfaces [11,12]. These capabilities have positioned membrane processes as key technologies for heavy metal remediation. The performance of polymeric membranes in water remediation is strongly influenced by the choice of polymer material. Commonly used polymers include polysulfone, polyethersulfone, polyvinylidene fluoride, cellulose acetate, and polyamide. Each material offers distinct advantages in terms of chemical resistance, mechanical strength, and permeability. However, challenges such as membrane fouling, limited selectivity, and trade-offs between permeability and rejection remain significant obstacles to widespread application.

https://doi.org/10.3390/membranes16030078

Membrane fouling, caused by the accumulation of organic matter, dyes, microorganisms, and inorganic salts on the membrane surface or within pores, is a major limitation in polymeric membrane processes. Fouling leads to reduced flux, increased energy consumption, and shortened membrane lifespan. To mitigate these issues, extensive research has focused on membrane surface modification strategies, including blending, coating, grafting, and incorporation of hydrophilic or antimicrobial additives. These approaches aim to enhance fouling resistance while maintaining or improving separation performance. Recent advances in polymer chemistry and nanotechnology have enabled the development of advanced polymeric membranes with enhanced functionality. The incorporation of nanoparticles, such as metal oxides, carbon-based nanomaterials, and zeolites, has led to the emergence of mixed-matrix and nanocomposite membranes. These materials can improve permeability, selectivity, mechanical stability, and contaminant affinity. Such innovations have shown particular promise in the removal of dyes and heavy metals from complex wastewater matrices [13,14]. Another important direction in membrane research is the design of membranes with tailored surface charge and functional groups. Functionalization with chelating agents, sulfonic groups, amines, or carboxyl groups can significantly enhance the binding and rejection of metal ions and charged dye molecules. These chemically modified membranes enable more selective and efficient removal processes, reducing the need for additional treatment steps and improving overall system sustainability. From a process perspective, membranes are often integrated into hybrid treatment systems to overcome the limitations of single-stage technologies. Coupling membrane filtration with adsorption, advanced oxidation processes or biological treatment can lead to synergistic effects, improving contaminant removal efficiency and operational stability [15,16]. Such integrated approaches are increasingly being explored for industrial wastewater treatment, where complex pollutant mixtures require multifunctional remediation strategies. Despite significant progress, several challenges must still be addressed to fully realize the potential of membranes in water remediation. Issues related to the long-term stability, fouling control, membrane recyclability, and environmental impact of membrane fabrication remain critical. Furthermore, the economic feasibility of large-scale applications depends on reducing material costs and energy consumption while maintaining high performance. Addressing these challenges requires interdisciplinary research combining materials science, chemical engineering, and environmental science.

Membranes have emerged as versatile and effective tools for the remediation of water contaminated with pollutants, dyes, and heavy metals. Their tunable properties, compatibility with advanced modification techniques, and suitability for integration into hybrid systems make them highly attractive for sustainable water treatment applications. Continued innovation in membrane materials and process design is expected to further enhance their performance and expand their role in addressing global water pollution challenges. This Special Issue (SI) aims to provide a comprehensive overview of recent developments and future perspectives in polymeric membrane technologies for water remediation membrane filtration and purification technologies. Membrane-based separation technologies continue to play a vital role in addressing global challenges related to water treatment, environmental protection, and sustainable energy systems. Ongoing advances in membrane materials, fabrication methods and process design are essential to improve separation efficiency, selectivity and long-term operational stability. This Special Issue, entitled "*Recent Advances in Membrane Filtration and Purification Technologies*", brings together eight original research articles that collectively reflect recent progress and emerging trends in membrane science and engineering.

Several contributions focus on pressure-driven membrane processes for water purification. Lim et al. (contribution 1) investigated the removal of phthalates using nanofil-

https://doi.org/10.3390/membranes16030078

tration and reverse osmosis membranes, emphasizing the influence of membrane surface properties on micropollutant rejection. Their findings provide important insights into membrane–solute interactions and offer guidance for the selection of membranes in advanced water treatment applications. In a relevant study, El Sayed et al. (contribution 2) evaluated a reverse osmosis-based zero liquid discharge system for the desalination of agricultural drainage water. By combining performance analysis with environmental and economic considerations, this work demonstrates the potential of membrane-based strategies to enhance water recovery while minimizing brine disposal. Advances in osmotically driven and energy-related membrane processes are also addressed. Moon and Kang (contribution 3) examined the effect of anion selection in tetrabutylphosphonium-based draw solutes on forward osmosis performance. Their results highlight the critical role of draw-solute chemistry in determining osmotic pressure and water flux, contributing to the optimization of forward osmosis systems.

In the context of hydrogen separation, Kashkarov et al. (contribution 4) examined Nb-Ni-Ti-Zr-Co high-entropy alloys as membrane materials. By correlating alloy microstructure with hydrogen permeability, the authors demonstrate the potential of high-entropy alloys for membrane applications in clean energy technologies.

A significant portion of this SI is devoted to the development of functional and nanocomposite membranes for pollutant removal. Alrebdi et al. (contribution 5) reported the fabrication of Ag-doped ZnO/MWCNT nanocomposite membranes via pulsed laser deposition, showing enhanced photocatalytic degradation of Rhodamine B. Similarly, Alrebdi et al. (contribution 6) developed Ag/ZnO thin-film nanocomposite membranes capable of efficiently degrading 4-nitrophenol, illustrating the advantages of integrating catalytic activity within membrane structures. In another contribution, Alkallas et al. (contribution 7) presented poly(vinyl alcohol)/Al_2O_3 nanocomposite films prepared by pulsed laser ablation in liquid, demonstrating effective removal of Ni(II) ions and highlighting the potential of nanocomposite membranes for heavy-metal remediation.

The Special Issue is brought to completion by a study on electrospun biodegradable composite membranes based on polyhydroxybutyrate and poly(vinyl formate) loaded with protonated montmorillonite performed by Penchev et al. (contribution 8). This work investigates the adsorption of organic dyes, including methylene blue and Congo red, and provides kinetic and isotherm analyses to elucidate the adsorption mechanisms. The results emphasize the role of fibrous membrane morphology and clay incorporation in enhancing dye removal performance, while also addressing sustainability considerations through the use of biodegradable polymers.

In summary, the eight articles collected within this SI highlight recent advances in membrane filtration and purification technologies, with particular emphasis on material innovation, multifunctional membrane design and application-oriented solutions. The contributions demonstrate the growing importance of membranes in water treatment, energy-related separations and environmental remediation. This Special Issue aims to serve as a valuable reference and stimulate further research in the field. As this SI draws to a close, it is clear that membrane science continues to advance at a remarkable pace, driven by the urgent need for more sustainable and effective separation technologies. The assembled contributions underscore significant progress in a broad range of membrane processes from traditional filtration and reverse osmosis to innovative hybrid and multifunctional systems. They highlight both fundamental insights and practical applications across water treatment, desalination, and industrial purification challenges. These studies reflect the community's commitment to not only deepening our understanding of membrane behavior but also addressing real-world issues such as energy use, fouling, and environmental impact. A consistent theme across the published works is the exploration of novel membrane materials and tailored structural designs that improve performance while mitigating long-standing obstacles like fouling and selectivity loss. Research articles in this issue have showcased material innovation, whether through hybrid composites, advanced surface functionalization, or engineered pore structures. They can deliver enhanced flux, selectivity, and durability under varied operational conditions. These advancements point toward a future in which membranes are not mere passive barriers but engineered interfaces capable of responsive and robust separation performance. The practical implications of these technological breakthroughs are equally noteworthy. Whether applied to wastewater recycling, oil/water emulsion separation, or protein purification, the cutting-edge studies presented here demonstrate the tangible benefits of membrane technology in tackling environmental and industrial purification tasks. Moreover, many contributions emphasize the balance between high separation efficacy and cost-effective design, aligning scientific innovation with the broader goals of sustainability and scalability in membrane applications.

Looking forward, the field is poised to make even greater strides by embracing interdisciplinary approaches that connect materials science, process engineering, and systems analysis. Continued collaboration among researchers, industry stakeholders, and policymakers will be essential to translate laboratory success into large-scale deployment that addresses global water scarcity, pollution, and resource recovery challenges. As this Special Issue has shown, the collective effort of the membrane community holds promise for not only refining existing technologies but also inspiring transformative solutions that contribute to a more sustainable future.

Acknowledgments: As Guest Editor of the Special Issue "Recent Advances in Membrane Filtration and Purification Technologies", I would like to express my deep appreciation to all contributors whose valuable works were published under this issue and thus participated in the success of the edition.

Conflicts of Interest: The author declares no conflicts of interest.

List of Contributions:

1. Lim, E.Q.; Seah, M.Q.; Lau, W.J.; Hasbullah, H.; Goh, P.S.; Ismail, A.F.; Emadzadeh, D. Evaluation of surface properties and separation performance of NF and RO membranes for phthalates removal. *Membranes* **2023**, *13*, 413. https://doi.org/10.3390/membranes13040413.
2. El Sayed, M.M.; Abulnour, A.M.; Tewfik, S.R.; Sorour, M.H.; Hani, H.A.; Shaalan, H.F. Reverse osmosis membrane zero liquid discharge for agriculture drainage water desalination: Technical, economic, and environmental assessment. *Membranes* **2022**, *12*, 923. https://doi.org/10.3390/membranes12100923.
3. Moon, J.; Kang, H. Anion effect on forward osmosis performance of tetrabutylphosphonium-based draw solute having a lower critical solution temperature. *Membranes* **2023**, *13*, 211. https://doi.org/10.3390/membranes13020211.
4. Kashkarov, E.; Krotkevich, D.; Koptsev, M.; Ognev, S.; Svyatkin, L.; Travitzky, N.; Lider, A. Microstructure and hydrogen permeability of Nb-Ni-Ti-Zr-Co high entropy alloys. *Membranes* **2022**, *12*, 1157. https://doi.org/10.3390/membranes12111157.
5. Alrebdi, T.A.; Rezk, R.A.; Alghamdi, S.M.; Ahmed, H.A.; Alkallas, F.H.; Pashameah, R.A.; Mostafa, A.M.; Mwafy, E.A. Photocatalytic performance improvement by doping Ag on ZnO/MWCNTs nanocomposite prepared with pulsed laser ablation method based photocatalysts degrading rhodamine B organic pollutant dye. *Membranes* **2022**, *12*, 877. https://doi.org/10.3390/membranes12090877.
6. Alrebdi, T.A.; Ahmed, H.A.; Alkallas, F.H.; Pashameah, R.A.; Alrefaee, S.H.; Alsubhe, E.; Mostafa, A.M.; Mwafy, E.A. Ag/ZnO thin film nanocomposite membrane prepared by laser-assisted method for catalytic degradation of 4-nitrophenol. *Membranes* **2022**, *12*, 732. https://doi.org/10.3390/membranes12080732.

7. Alkallas, F.H.; Ahmed, H.A.; Alrebdi, T.A.; Pashameah, R.A.; Alrefaee, S.H.; Alsubhe, E.; Trabelsi, A.B.G.; Mwafy, E.A. Removal of Ni (II) ions by poly (Vinyl Alcohol)/Al_2O_3 nanocomposite film via laser ablation in liquid. *Membranes* **2022**, *12*, 660. https://doi.org/10.3390/membranes12070660.
8. Penchev, H.; Abdelhamid, A.E.; Ali, E.A.; Budurova, D.; Grancharov, G.; Ublekov, F.; Koseva, N.; Zaharieva K.; El-Sayed, A.A.; Khalil, A.M. Novel electrospun composite membranes based on polyhydroxybutyrate and poly (vinyl formate) loaded with protonated montmorillonite for organic dye removal: Kinetic and isotherm studies. *Membranes* **2023**, *13*, 582. https://doi.org/10.3390/membranes13060582.

References

1. Bernardo, P.; Drioli, E.; Golemme, G. Membrane gas separation: A review/state of the art. *Ind. Eng. Chem. Res.* **2009**, *48* 4638–4663. [CrossRef]
2. Wang, Y.; Ghanem, B.S.; Ali, Z.; Hazazi, K.; Han, Y.; Pinnau, I. Recent progress on polymers of intrinsic microporosity and thermally modified analogue materials for membrane-based fluid separations. *Small Struct.* **2021**, *2*, 2100049. [CrossRef]
3. Khosla, A.; Sonu; Awan, H.T.A.; Singh, K.; Gaurav; Walvekar, R.; Zhao, Z.; Kaushik, A.; Khalid, M.; Chaudhary, V. Emergence of MXene and MXene–polymer hybrid membranes as future-environmental remediation strategies. *Adv. Sci.* **2022**, *9*, 2203527 [CrossRef] [PubMed]
4. Werber, J.R.; Deshmukh, A.; Elimelech, M. The critical need for increased selectivity, not increased water permeability, for desalination membranes. *Environ. Sci. Technol. Lett.* **2016**, *3*, 112–120. [CrossRef]
5. Reid, C.; Liu, W.; Aguey-Zinsou, K.F. Metal Hydrides-Based Hydrogen Purification Systems—A Review. *MetalMat* **2025**, e70025 [CrossRef]
6. Karousos, D.S.; Qadir, D.; Sapalidis, A.A.; Ahmad, F.; Favvas, E.P. Polymeric, metallic and carbon membranes for hydrogen separation: A review. *Gas Sci. Eng.* **2023**, *120*, 205167. [CrossRef]
7. Santhamoorthy, M.; Asaithambi, P.; Perumal, I.; Elangovan, N.; Natarajan, P.; Lin, M.C.; Kim, S.; Kumarasamy, K.; Phan, T.T.V. A comprehensive review of the functionalized polymer composite membranes in wastewater treatment. *J. Environ. Chem. Eng.* **2025**, *13*, 117735. [CrossRef]
8. Noah, N.M. Current status and advancement of nanomaterials within polymeric membranes for water purification. *ACS Appl. Nano Mater.* **2023**, *7*, 18610–18625. [CrossRef]
9. Penchev, H.; Abdelhamid, A.E.; Ali, E.A.; Budurova, D.; Grancharov, G.; Ublekov, F.; Koseva, N.; Zaharieva, K.; El-Sayed, A.A.; Khalil, A.M. Novel electrospun composite membranes based on polyhydroxybutyrate and poly (vinyl formate) loaded with protonated montmorillonite for organic dye removal: Kinetic and isotherm studies. *Membranes* **2023**, *13*, 582. [CrossRef] [PubMed]
10. Bashir, N.; Afzaal, M.; Khan, A.L.; Nawaz, R.; Irfan, A.; Almaary, K.S.; Dabiellil, F.; Bourhia, M.; Ahmed, Z. Green-synthesized silver nanoparticle-enhanced nanofiltration mixed matrix membranes for high-performance water purification. *Sci. Rep.* **2025**, *15*, 1001. [CrossRef] [PubMed]
11. Nayak, D.T.; Raja, V.K.; Arthanareeswaran, G.; Khoa, T.D.; Taweepreda, W. A high-performance nanofiltration membrane synthesized by embedding amino acids and ionic liquids in cellulose acetate for heavy metal separation. *RSC Sustain.* **2025**, *3*, 1966–1981. [CrossRef]
12. Penchev, H.; Budurova, D.; Ublekov, F.; Zaharieva, K.; Ali, E.A.; Abdelhamid, A.E.; El-Sayed, A.A.; Khalil, A.M. Formic acid electrospun composite fibers based on polyhydroxybutyrate/polyamide 6 loaded with protonated montmorillonite as promising disposable heavy metal removal sorbent. *Emerg. Mater.* **2025**, *8*, 2451–2462. [CrossRef]
13. Elfiky, A.A.E.A.; Mubarak, M.F.; Keshawy, M.; Sayed, I.E.T.E.; Moghny, T.A. Novel nanofiltration membrane modified by metal oxide nanocomposite for dyes removal from wastewater. *Environ. Dev. Sustain.* **2024**, *26*, 19935–19957. [CrossRef]
14. Lustenberger, S.; Castro-Muñoz, R. Advanced biomaterials and alternatives tailored as membranes for water treatment and the latest innovative European water remediation projects: A review. *Case Stud. Chem. Environ. Eng.* **2022**, *5*, 100205. [CrossRef]
15. Wang, Z.; Zhang, Y.; Li, K.; Guo, J.; Yang, C.; Liu, H.; Wang, J. In situ coupling of electrochemical oxidation and membrane filtration processes for simultaneous decontamination and membrane fouling mitigation. *Sep. Purif. Technol.* **2022**, *290*, 120918. [CrossRef]
16. Singh, J.; Saharan, V.; Kumar, S.; Gulati, P.; Kapoor, R.K. Laccase grafted membranes for advanced water filtration systems: A green approach to water purification technology. *Crit. Rev. Biotechnol.* **2018**, *38*, 883–901. [CrossRef] [PubMed]

https://doi.org/10.3390/membranes16030078

membranes

MDPI

Article

Novel Electrospun Composite Membranes Based on Polyhydroxybutyrate and Poly(vinyl formate) Loaded with Protonated Montmorillonite for Organic Dye Removal: Kinetic and Isotherm Studies

Hristo Penchev [1,*], Ahmed E. Abdelhamid [2], Eman A. Ali [2], Dessislava Budurova [1], Georgy Grancharov [1], Filip Ublekov [1], Neli Koseva [1], Katerina Zaharieva [3], Ahmed A. El-Sayed [4] and Ahmed M. Khalil [4,*]

1 Institute of Polymers, Bulgarian Academy of Sciences, Acad. G. Bonchev 103A, 1113 Sofia, Bulgaria; dbudurova@gmail.com (D.B.); granchar@polymer.bas.bg (G.G.); ublekov.philip@gmail.com (F.U.); koseva@polymer.bas.bg (N.K.)

2 Polymers and Pigments Department, National Research Centre, El-Buhouth Str., Giza 12622, Egypt; ahmednrc10@gmail.com (A.E.A.); emabal2003@yahoo.com (E.A.A.)

3 Institute of Mineralogy and Crystallography Acad. I. Kostov, Bulgarian Academy of Sciences, Acad. G. Bonchev 107, 1113 Sofia, Bulgaria; zaharieva@imc.bas.bg

4 Photochemistry Department, National Research Centre, El-Buhouth Str., Giza 12622, Egypt; ahmedcheme4@yahoo.com

* Correspondence: hpen4ev@gmail.com (H.P.); akhalil75@yahoo.com (A.M.K.)

Abstract: The use of biodegradable polyesters derived from green sources and their combination with natural abundantly layered aluminosilicate clay, e.g., natural montmorillonite, meets the requirements for the development of new sustainable, disposable, and biodegradable organic dye sorbent materials. In this regard, novel electrospun composite fibers, based on poly β-hydroxybutyrate (PHB) and in situ synthesized poly(vinyl formate) (PVF), loaded with protonated montmorillonite (MMT-H) were prepared via electrospinning in the presence of formic acid, a volatile solvent for polymers and a protonating agent for the pristine MMT-Na. The morphology and structure of electrospun composite fibers were investigated through SEM, TEM, AFM, FT-IR, and XRD analyses. The contact angle (CA) measurements showed increased hydrophilicity of the composite fibers incorporated with MMT-H. The electrospun fibrous mats were evaluated as membranes for removing cationic (methylene blue) and anionic (Congo red) dyes. PHB/MMT 20% and PVF/MMT 30% showed significant performance in dye removal compared with the other matrices. PHB/MMT 20% was the best electrospun mat for adsorbing Congo red. The PVF/MMT 30% fibrous membrane exhibited the optimum activity for the adsorption of methylene blue and Congo red dyes.

Keywords: electrospinning; nanocomposite fibers; polyhyrdoxybuterate; poly(vinyl formate); formic acid; protonated montmorillonite; dye removal; methylene blue; Congo red

1. Introduction

During the past decade, the study of biodegradable and bio-based polyesters has gained great popularity, not exclusively for research purposes but also for industrial applications. A distinguished member of the polyhydroxyalkanoate polyester family is poly β-hydroxybutyrate (PHB). It is obtained through multi-step extraction/purification from different bacterial strains or through synthetic pathways [1–4]. It is favored for being eco-friendly, biodegradable, and compatible with other thermoplastics. Well-known PHB disadvantages include its hydrophobicity, its high degree of crystallinity and inherent brittleness, and its low surface functionality. In this regard, the emerging and already industrially applicable electrospinning technique may offer a convenient solution for acquiring PHB with increased hydrophilicity and functionality upon blending it with polar groups containing polymers, e.g., poly(ethylene oxide) [5] or poly(vinyl alcohol), PVA [6]. The latter is especially desirable, as commercial PVA co-polymers are cheap, industrially available, and rich in functional polar -OH and acetate groups, allowing for further chemical modifications or the compatibilization of different metal oxides and clay nanoparticles [7]. In this regard, PHB/PVA-blend fibrous mats were obtained with the use of volatile perfluorinated alcohol—hexafluoro-2-propanol as a common polymer blend solvent, producing electrospun mats with reduced crystallinity and improving their polarity and hydrophilicity [8]. The polar nature of PVA requires the use of expensive and environmentally/health hazardous PHB compatibilization solvents instead of the traditionally used, cheap, and less environmentally hazardous chlorinated hydrocarbons, e.g., chloroform and dichloromethane, which unfortunately act as anti-solvents for PVA. In this regard, the recent work of Penchev et al. showed the use of 99% formic acid (FA) as an innovative green chemistry solvent for electrospinning pure PHB and its blends with a widely available plant-extracted polypeptide: soy protein. This work also represents a great potential for other polar–nonpolar polymer–inorganic nanoparticle hybrid blends and their compatibilization and subsequent electrospinning [9]. Another known, but rarely explored, feature of FA is its ability to induce the esterification of dissolved PVA in situ under mild conditions with a produced water-insoluble poly(vinyl formate) (PVF) co-polymer [10,11].

Among natural clay materials, sodium montmorillonite (MMT-Na) is a well-known and widely explored material that is used as an effective layered aluminosilicate sorbent for aqueous organic dye removal and water remediation; it is also a polymer reinforcing filler [12–14]. This clay exhibits a high specific surface area that interacts with polymeric chains to boost their properties in resulting nanocomposites. Montmorillonite (MMT) has attracted considerable attention as a mineral clay in the field of flake-based hybrids. This is on account of its peculiar features, such as swelling with a high lamellar expansion capacity in aqueous solutions [15]. Montmorillonite is one of the most widely used layered aluminum silicate clays with exchangeable cations and reactive –OH groups on its surface [16]. It is formed in layers, maintaining a platelet structure. The interlayer spacing is nearly 1 nm [17]. Pristine montmorillonite exhibits a hydrophilic nature. The polar nature of MMT-Na is the reason for its incompatibility with non-polar hydrophobic polymers and polyesters, in particular, those without surface hydrophobization treatment and the resultant organo-modified montmorillonites (OMMT), which are intrinsically compatible with polyester matrices. However, the incorporation of OMMT into hydrophobic polymer matrices is a particular and a commonly applied approach in polyester nanocomposite preparation. It does not solve the problem of imparting the needed hydrophilic-hydrophobic balance and increased water sorption capacity desired in manufacturing disposable substrate composite materials for aqueous dye removal applications. Nanoclay-polymer composites exist in either intercalated or exfoliated forms. Some polymeric chains disperse into the nanofiller layers with a steady interlayer spacing, forming intercalated composites [18].

Organic dyes are widely used in the coating industry, the textile industry, the photocatalytic industry, and photochemical applications. Treatment of contaminated and colored wastewater from textile or other industries is an important problem that has attracted researchers' attention [19]. Different pathways have been followed to remove dyes from aqueous media. They include ion-exchange, membrane separation, and adsorption. Adsorption is considered a facile method because it is easily implemented and produces nontoxic byproducts. As mentioned, aluminosilicate clays such as kaolinite and montmorillonite are frequently used for dye removal applications. However, collecting and abstracting the accumulated dye sorbent in the form of fine dispersion is still a challenge for current technology [20–22]. Methylene blue is a typical cationic dye. Various studies have investigated how to remove it from water through adsorption or photochemical decomposition [23,24]. Researchers found that montmorillonite clay is effective as an ion-exchange type adsorbent for the removal of both cationic methylene blue and anionic Congo red dyes [25,26]. Yusef Khaniabadi et al. investigated natural montmorillonite as a low-cost and effective absorbent for Congo red sorption from aqueous solutions and showed that the data were well fitted by pseudo-second order kinetic and Freundlich isotherm models [27]. Another report showed that the removal of acid dyes was improved using sulfuric acid-activated bentonite clay compared with untreated bentonite. They showed that the protonated clays displayed higher adsorption capacities than the natural bentonite clay [28].

Based on our experience and accumulated knowledge, we have designed the present study with the aim of preparing spinnable highly protonated montmorillonite composites based on PHB or PVF, main-chain and side-chain polyesters, respectively. These composites are produced via solution dispersion electrospinning. This new approach could be a promising green chemistry pathway for the use of formic acid as common polymer solvent instead of perfluorinated hydrocarbon solvent alternatives, and its simultaneous use as an efficient dispersant of MMT-H as well. We have found that during the dispersion process, formic acid reacts instantly with natural MMT-Na forms, transforming them into protonated MMT-H forms possessing strong dye sorption properties that act as reactive solvents for the in situ esterification of PVA into PVF co-polymers. The result was preparation of well-dispersed electrospun composite fibrous membranes based on PVF/MMT-H and/or PHB/MMT-H with cationic (methylene blue) and anionic (Congo red) dyes sorption properties. These properties were studied in detail through kinetics and isotherm investigations.

2. Materials and Methods

2.1. Materials Characterization

Poly β-hydroxybutyrate powder from a bacterial origin with an average molecular mass of 230,000 g·mol^{-1} was kindly supplied by Dr. Urs Hanggi (Biomer, Krailling, Germany). Poly(vinyl alcohol) fine powder with a DH of 97.5–99.5% and a Mw avr. = 72,000 g·mol^{-1} from Merck-Schuchardt (Hohenbrunn, Germany)) was used. Pure formic acid (Merck, 99%) was used as a solvent. Natural montmorillonite clay (MMT-Na) was supplied by Southern Clay Products Inc, USA. For electrospinning experiments, a commercial Genvolt HV PSU 30 W direct current power supply of 0–30 kV and an Alaris IVAC® P3000 medical syringe pump were used. Wide-angle X-ray diffraction (WAXD) scans were obtained with a Bruker D8 Advance ECO diffractometer in reflection mode over a 2θ range of 5–50°. The ATR FT-IR analyses were performed on a Thermo Scientific Nicolet iS50 apparatus in the range of 400–4000 cm^{-1}. Thermogravimetric analyses were performed in an argon atmosphere on a TGA-4000 Perkin Elmer instrument (temperature range 30 °C–800 °C, 10 °C/min, nitrogen atmosphere, 30 mL·cm^{-1}). SEM analysis was performed on a high-resolution scanning electron microscope (SEM) (QUANTA FEG 250 ESEM). TEM analyses were carried out on an HR STEM JEOL JEM 2100. AFM imaging was performed on a Bruker Dimension Icon microscope working under peak force tapping mode. The statistical measurements of fibers were performed with Image J software. The contact angle values were recorded with a camera attached to a computer. Deionized water droplets (5 μL) were transferred to the surface of the membrane at 25 °C to explore the membrane hydrophilicity.

2.2. Preparation of PVF- and PHB/MMT-H Spinning Dispersions

In order to prepare electrospinable MMT dispersion solutions, the first step includes preparing a stock-spinning solution of in situ-formed PVF. This was performed by dissolving 0.5 g of PVA powder in 9.5 g FA at 60 °C for 3 h under constant stirring. The obtained viscous solution was left to stay at room temperature (25 °C) for 24 h before electrospinning experiments in order to reach a high degree of esterification. The degree of esterification (DE) for the obtained PVF co-polymer was determined gravimetrically by casting a known amount of the resultant PVF solution with a doctor's blade (0.1 mm gap) onto a glass substrate. Sol-gel immersion of the film in a water bath and subsequent vacuum oven drying were carried out for the inversed gel film at 70 °C overnight. The experimentally determined DE of the pristine PVA was 80%. The spinning dispersion of PVF with 30 wt.%

MMT content was performed by dispersing 0.25 g MMT-Na in 9.5 g FA using a laboratory vortex followed by an additional 10 min ultrasound bath treatment. The obtained pristine in situ-formed protonated montmorillonite clay (MMT-H) was further sterically stabilized by dissolving 0.5 g PVA according to the described procedure. The final milky white fine dispersion was left overnight under constant stirring before electrospinning. Stock dispersions of MMT-H (10 and 20 wt.% on polymer weight base) in PHB were prepared by adding and sonicating 0.05 g (for PHB/MMT 10% composite) or 0.1 g (for PHB/MMT 20%) of pristine MMT-Na in 7 g FA. This step was followed by the addition of 0.5 g PHB powder after 5 min of ultrasound treatment until a homogeneous dispersion of both components took place.

Thereafter, the common particle dispersion of MMT/PHB was heated in a silicon oil bath at 80 °C for 10 min under magnetic stirring until the PHB particles were fully dissolved. The resulting MMT dispersion in PHB was left to cool down to 40 °C before electrospinning. The control pristine PHB electrospinning solution was prepared in the same manner without the addition of MMT-Na. The PHB/PVF/MMT 10% (component weight ratio: 63:27:10%) spinning solution was prepared by mixing and homogenization 3 g of PVF/MMT 30 stock dispersion with 10 g of 5 wt.% solution of PHB in FA at 60 °C. After obtaining a homogeneous mixture, the spinning solution was left to cool down to 40 °C before the electrospinning experiment.

2.3. Electrospinning Process

The electrospinning experiments were carried out with a fixed applied voltage of 20 kV and a stationary aluminum collector (20 × 20 cm, 2 mm thickness). The spinning solution flow rates and the distance between the electrodes were varied, depending on the electrospun sample polymer/MMT composition, as follows: PVF-0: 0.2 $mL \cdot h^{-1}$, 15 cm; PVF/MMT 30%—0.5 $mL \cdot h^{-1}$, 15 cm; PHB/MMT 10%—0.7 $mL \cdot h^{-1}$, 20 cm; PHB/MMT 20%—0.5 $mL \cdot h^{-1}$, 20 cm; and PHB/PVF/MMT 10%—1.5 $mL \cdot h^{-1}$, 15 cm. The composite mats were immersed for 24 h in water to remove the residual sodium formate salt then dried in a vacuum chamber. In Table 1, the compositions of different samples and their electrospinning parameters are summarized.

Table 1. The prepared samples' compositions and spinning conditions.

	Composition (wt.%)			Electrospinning Parameters		
Sample Name	**PHB**	**PVF**	**MMT**	**Flow Rate ($mL \cdot h^{-1}$)**	**Voltage (kV)**	**Distance to Collector (cm)**
PVF	0	100	0	0.2	20	15
PVF/MMT 30%	0	70	30	0.5	20	15
PHB	100	0	0	0.5	20	20
PHB/MMT 10%	90	0	10	0.7	20	20
PHB/MMT 20%	80	0	20	0.5	20	20
PHB/PVF/MMT 10%	63	27	10	1.5	20	15

2.4. Dye Sorption and Kinetics Studies of the Composite Membranes

The adsorption experiments of cationic dye (methylene blue) and anionic dye (Congo red) for the investigated membranes were conducted at room temperature (25 ± 2 °C) under different contact time intervals (0–240 min). The initial dye concentration ranged from 10–50 mg/L. The adsorption experiment was designed as a batch in stoppered reagent bottles at a constant dose of 1 g/L. The shaking speed was 150 rpm. The concentrations of methylene blue and Congo red before and after adsorption were determined using a UV-Visible Spectrophotometer (Cary 100 UV-Vis) at wavelengths of 665 and 498 nm, respectively.

The adsorption capacity (qe in mg/g) was calculated by the following equation:

$$qe = \frac{Co - Ce}{m} * V \quad (1)$$

where *Co* and *Ce* are the initial and final dye concentrations (mg/L), *V* is volume of dye solution (L), and m is the weight of the adsorbent fiber mats (g). This study examined the interactions between pollutants (dyes) and the nanofiber membranes using adsorption isotherm models. The Freundlich and Langmuir models are the two adsorption isotherm models which are commonly used. In order to identify the best model for our purposes, linear regressions of Langmuir (Equation (2)) and Freundlich (Equation (3)) models were used.

$$C_e/q_e = C_e/q_{max} + 1/(q_{max}K_L) \quad (2)$$

$$\ln q_e = \frac{1}{n}\ln C_e + \ln K_F \quad (3)$$

where C_e is the equilibrium concentration of the dye in the solution (mg/L) and q_e stands for the adsorption capacity at equilibrium (mg/g). q_e was calculated from triplicate measurements for three separate replicates of each experiment. q_{max} is the maximum adsorption capacity (mg/g), K_L and K_F are the adsorption rate constants of the Langmuir and Freundlich models, respectively, and n is the heterogeneity parameter [29]. Using linear regression, correlation coefficients (R^2) were estimated and the adsorption isotherm models were evaluated for applicability.

Adsorption Kinetics

Two kinetic models, pseudo-first order and pseudo-second order equations, were used to assess the rate of dye adsorption in the prepared PVA nanocomposites mats [30–32]. The pseudo-first order equation was represented as follows:

$$Ln\ (qe - qt) = Ln\ qe - k1\ t \quad (4)$$

where *k*1 (1/min) is the rate constant per minute and *qe* and *qt* (mg/g) are the amount of adsorbed dye at equilibrium and at time *t* (min), respectively.

The pseudo-second order equation was represented as follows:

$$\frac{t}{qt} = \frac{t}{k2qe^2} + \frac{t}{qe} \quad (5)$$

where *k*2 (g/mg.min) is the pseudo-second order rate constant and *qe* (mg/g) is the quantity of adsorbed dye at equilibrium.

3. Results and Discussion

3.1. Characterization of the Electrospun Membranes

In the present study, we explored the ability of the volatile and highly polar 99% formic acid to disperse natural montmorillonite (MMT-Na) effectively and convert it to protonated MMT-H in situ. Hence, an ion-exchange reaction takes place with the formation of sodium formate salt as a byproduct. Moreover, FA acts as a highly efficient reactive solvent for the in situ esterification of water-soluble pristine PVA co-polymers. It also tends to be water insoluble as an amphiphilic poly(vinyl formate) co-polyester. These two general properties led us to the idea to use FA as a green chemistry alternative to the state of the art perfluorinated low-alcohols and the trifluoracetic acid solvents for the electrospinning of PVF/MMT 10%, PHB/MMT 20%, PHB/MMT 30%, and the PHB/PVF/MMT (70:30/10 wt.%) composites, as illustrated through the general materials preparation scheme in Figure 1. The latest composite will be further referred to as PHB/PVF/MMT 10%.

Compatibilization of polar MMT-Na with the use of traditional aprotic solvents (e.g., chloroform and dichloromethane) for biodegradable polyester (e.g., PHB) dissolution is a challenging task. The most commonly used approach in this case is known from the literature on polyester compatibilization and involves blending with hydrophobic organo-modified montmorillonites [33]. However, with this common approach, the resultant composites retain, or even increase, their hydrophobicity. Thus, they could be considered mostly unsuitable as aqueous media dyes and heavy metal ion removal sorbents. In this regard, the works of Iordanskii et al. and Penchev et al. exploring the use of pure formic acid or its mixture with chloroform as green chemistry polar (co-)solvents demonstrated their applicability for electrospinning of pure PHB or PHB blends with soy protein [9,34]. These pioneering works were the base for further research into the possible compatibilization of natural MMT-Na, which has been already mentioned, and effective aqueous dye removal sorbents. This feature is due to MMT-Na's layered platelet structure and ion-exchange properties [12]. The immobilization of MMT in high specific surface electrospun biodegradable fibrous matrices is beneficial for the safe removal of sorbed organic dye contaminants from the environment and for the utilization of nanoclay composite membranes in disposable biodegradable forms. In this regard, our experiments started with the use of pure formic acid (FA) solvent, which acts as both a protonating agent for converting sodium MMT to protonated MMT-H and as a dispersion media. Simultaneously, we used FA as a reactive solvent for the in situ esterification of pristine water-soluble PVA to water-insoluble side-chain polyester PVF, which disperses commonly with MMT-H. Thereafter, FA sterically stabilizes the dispersed clay particles. We found that the obtained PVF solution can effectively stabilize up to 30 wt.% MMT-H clay. The resulting highly concentrated stock solution can be used for the preparation of highly loaded polymers with MMT-H composite fibers (PVF/MMT 30%). Moreover, it may be mixed with PHB solution to obtain polymer composite fibers with lower MMT-H contents (PHB/PVF/MMT 10%) via electrospinning. Furthermore, two different pure PHB composite fibers with 10% and 20 wt.% MMT-H were incorporated in the preparation from formic acid solution. The lower solution viscosity of PHB and its main-chain polyester nature can lead to a maximum loading of MMT-H in which stable electrospinning occurs upon loading 20 wt.% clay.

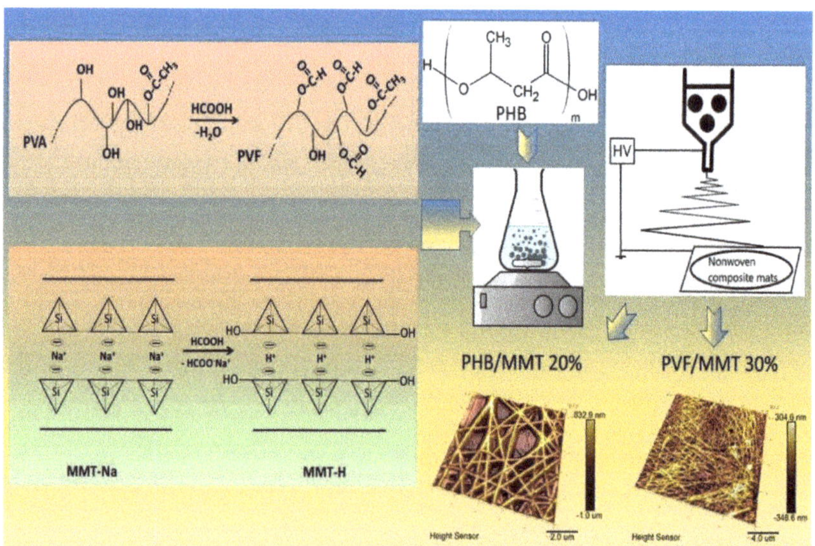

Figure 1. General materials preparation scheme of the novel dye sorption electrospun membranes.

In Figure 2, SEM micrographs of the electrospun mats are presented. The average fiber diameters determined via SEM are as follows: 139 nm ± 53 (PVF); 157 nm ± 119 (PVF/MMT 30%); 265 nm ± 105 (PHB); 267 nm ± 161 (PHB/MMT 10%); and 304 nm ± 109 (PHB/PVF/MMT 10%). As can be seen, the morphology of the composite fibers is affected by the incorporation of proton-conducting MMT-H. There is an observed increase in the average fiber's diameter with the formation of more defects and branched structures. However, adding the second side-chain polyester component, PVF, led to a significant increase in the smoothness and homogeneity of the PHB/PVF/MMT 10% fibers compared with the PHB/MMT 10% and PHB/MMT 20% composite fibers, which, as mentioned, showed some bead defects and partially intercrossing fibrous structures. This could be explained by the more flexible chain nature of PVF co-polymers and some additional intermolecular wrapping between PHB and PVF macromolecules in FA solution.

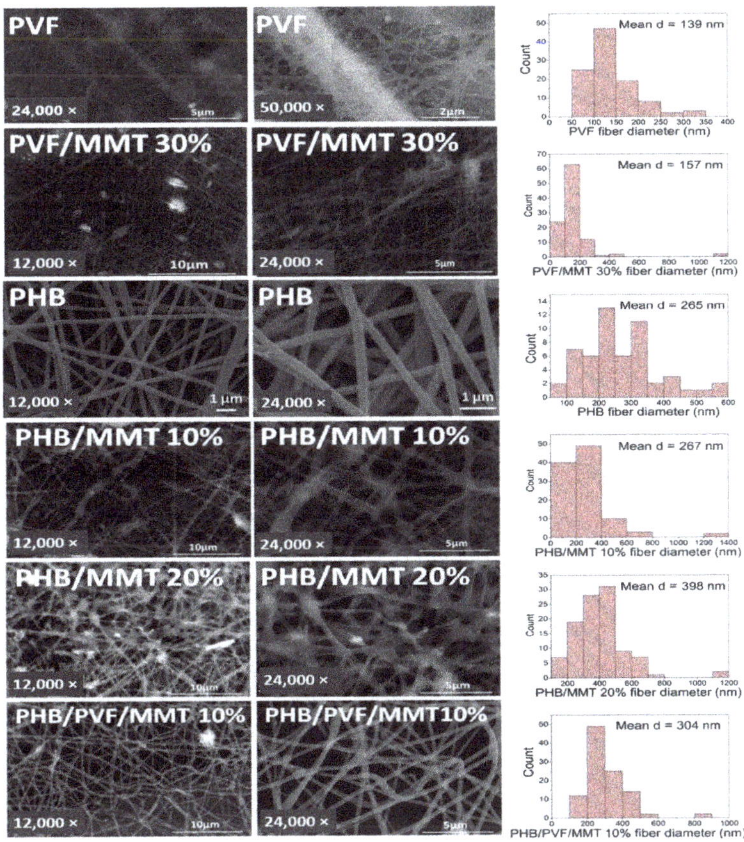

Figure 2. SEM images of PVF, PVF/MMT 30%, neat PHB, PHB/MMT 10%, PHB/MMT 20%, and PHB/PVF/MMT 10%, with the average diameter distribution of these fibers.

The TEM analyses of the investigated membranes are depicted in Figure 3. The composite PHB/MMT 10% fibers in Figure 3A show that mostly exfoliated MMT nanosheets are incorporated into the fibrous polyester matrix. In the case of composite PHB/PVF/MMT fibers, some nanophase clusters are distributed along the fibers' axes (Figure 3B). Figure 3C shows that the TEM micrographs of the PHB/PVF/MMT mat after 20 min in contact with aqueous 1 mg/L methylene blue solution revealed a swelling in the composite fibers. Moreover, the presence of expanded hydrated MMT clusters along the fiber axis is proof of their good sorption property.

The 2-D and 3-D AFM surface topography images of the PHB/MMT 20% and PVF/MMT 30% composite electrospun fibrous mats are shown in Figure 4A–D. As can be seen, the PHB/MMT 20% mat forms a loose, smooth-surfaced fibrous membrane structure (Figure 4A,B) compared with the very dense, small-diameter, well-overlapped fibrous structure of PVF/MMT 30% composite membranes (Figure 4C,D). The AFM surface topography images of the MMT-H-containing composite fibrous membranes after 10 h of contact with aqueous 1 mg/L methylene blue solution showed significant changes in both samples (Figure 5A–D). A noticeable swelling of the fibers is observed after contact with the aqueous dye solution, which is in accordance with the TEM observation. The 2D and 3D surface morphology of the PHB/MMT 20% mat after water contact revealed swelling of the polymer matrix and less sharp fiber contrast morphology compared with the dry pristine mat (Figures 4 and 5A,B). In both dye-adsorbed composite polyester fiber samples, some random sections of expanded MMT particles over the fiber surface are visible. This could be confirmation of the already observed results of the TEM analysis of MMT platelet clustering.

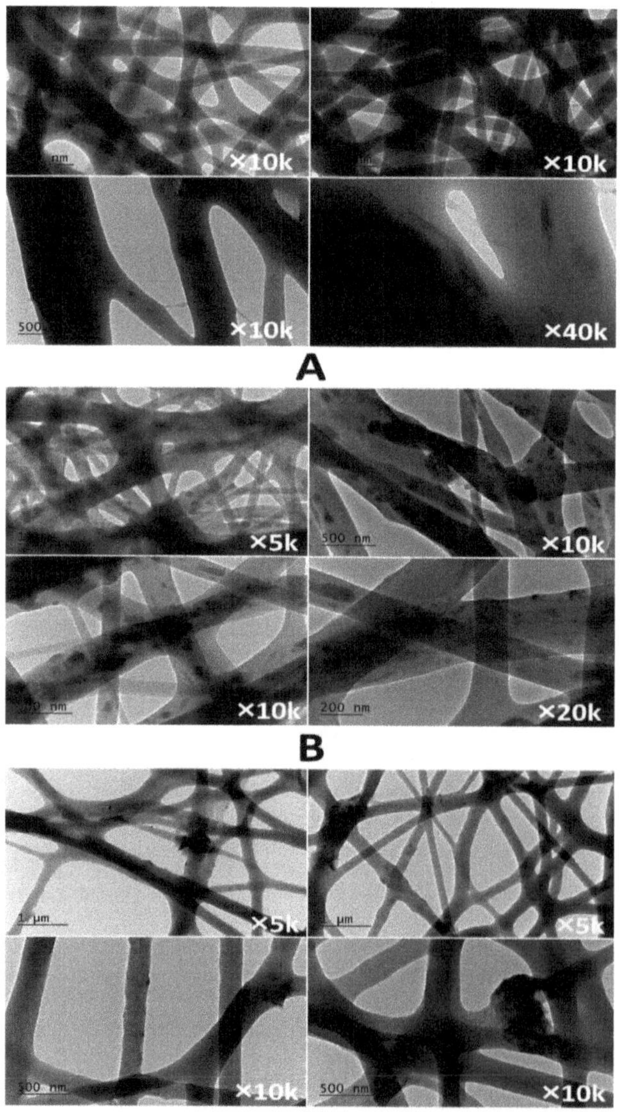

Figure 3. TEM images of (**A**) PHB/MMT 10%; (**B**) PHB/PVF/MMT 10%; (**C**) PHB/PVF/MMT 10% after aq. MB sorption.

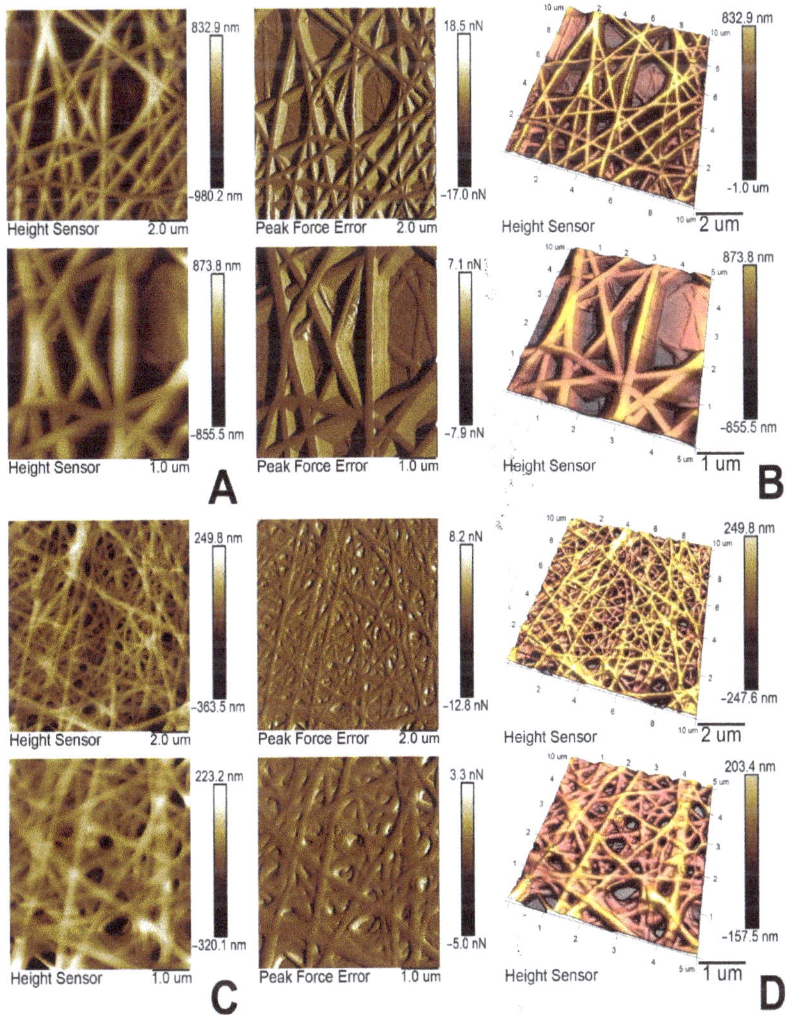

Figure 4. (**A**) AFM 2-D images of PHB/MMT 20%; (**B**) AFM 3-D images of PHB/MMT 20%; (**C**) AFM 2-D images of PVF/MMT 30%; and (**D**) AFM 3-D images of PVF/MMT 30%.

The TGA and DTG curves of the starting materials and the composite electrospun polyester fibers with incorporated MMT-H are clarified in Figure 6. Three stages are registered during the thermal degradation of both pristine PVA and the in situ esterified mats by the formic acid solvent PVF co-polymer. The lower mass loss (1.81%) of PVF below 200 °C in comparison with that of pristine PVA is due to an elevated hydrophobicity. It results from the FA esterification of the hydrophilic -OH groups of PVA. The first weight loss (4.17%) of PVA below 200 °C was attributed to the evaporation of free water molecules in PVA due to its hydrophilic property [35]. The second step of decomposition is attributed to main chain degradation and the generation of CO, CO_2, and volatile hydrocarbons from the PVF matrix. The third step of degradation of PVF is assigned to the carbonization process, with a small fraction of char formation. The increased thermal stability of PVF/MMT 30% and the slightly increased stability of the PHB/MMT 10% electrospun composite membranes compared to pristine PVF and PHB fibers may be due to the incorporation of the well-dispersed MMT-H clay in the polymer matrix of electrospun mats. This effect is most profound in the PVF/MMT 30% sample. MMT improves the structural stability of polymer fibers [36]. The resulting residue above the degradation temperature of the polymers is associated with the presence of thermally intact montmorillonite and a small fraction of charred material.

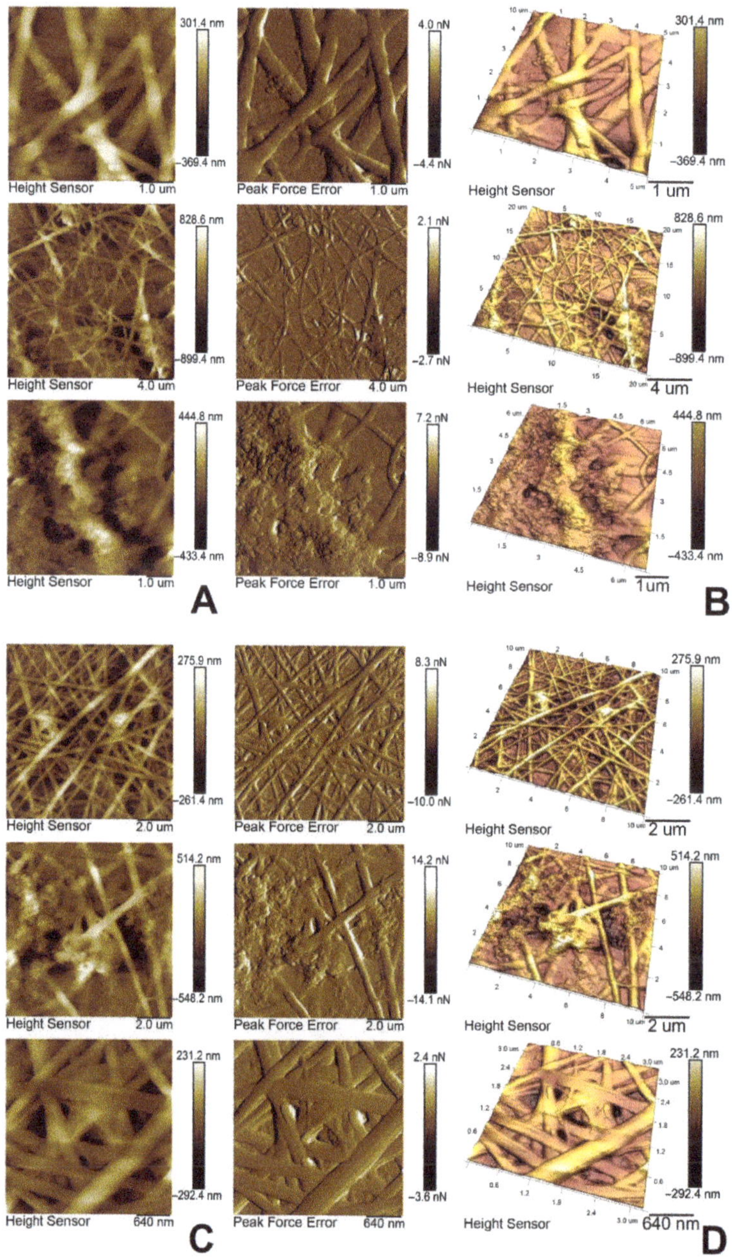

Figure 5. (**A**) AFM 2-D images of PHB/MMT 20%; (**B**) AFM 3-D images of PHB/MMT 20%; (**C**) AFM 2-D images of PVF/MMT 30%; and (**D**) AFM 3-D images of PVF/MMT 30% after MB sorption.

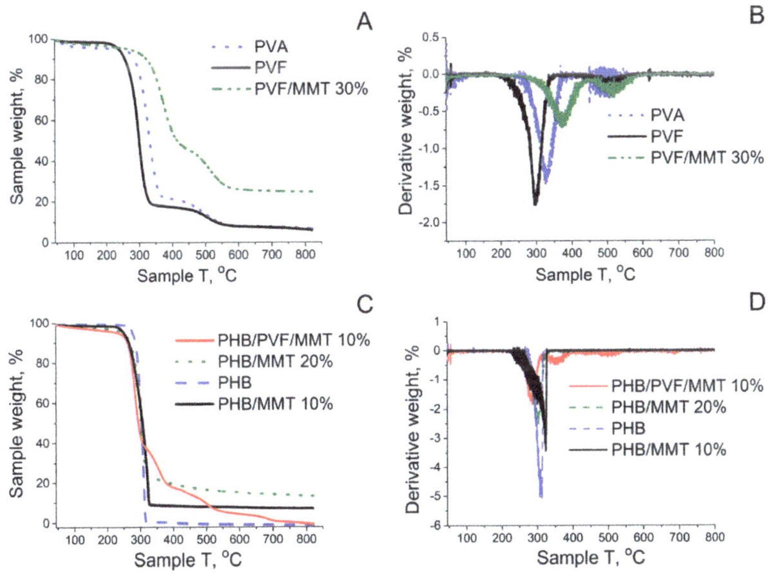

Figure 6. (**A**) TG and (**B**) DTG curves of PVA, PVF, and PVF/MMT 30%; (**C**) TG and (**D**) DTG curves of PHB, PHB/MMT 10%, PHB/MMT 20%, and PHB/PVF/MMT 10%.

The XRD analyses of pristine MMT-Na and the in situ protonated MMT-H form obtained after the ion-exchange reaction with formic acid during the dispersion process are presented in Figure 7A. The structure of the nanocomposites is demonstrated by the enlargement degree of d001-spacing. The protonation of MMT-Na in the presence of formic acid shifts the d001 to lower angles, from $2\theta = 7.88°$ (d001 = 11.21 nm) in MMT-Na to $2\theta = 7.02°$ (d00 = 12.58 nm). The XRD patterns of electrospun composite fibers, namely PHB/MMT 10% and PHB/PVF/MMT 10%, show a broad amorphous halo with two main diffraction peaks at 13.4° and 16.8°. They correspond to the orthorhombic crystal structure of PHB (Figure 7B). The XRD patterns of PHB/PVF/MMT 10% and PHB/MMT 10% indicate an exfoliation of MMT-H platelets for which there is a lack of a d001 diffraction peak. In order to prove further the existence of exfoliated MMT polymer composites obtained during the electrospinning process, we performed an experiment in which we dissolved a free-standing fibrous mat sample of PHB/MMT 10% in polar aprotic solvent (chloroform). After contact between the mat and the chloroform, almost immediate dissolution of the PHB matrix (2 wt. %) takes place. The obtained solution has a milky opalescence, which is an indication of the colloidal dispersion of the MMT-H in the PHB/chloroform solution [37]. After that, the chloroform dispersion was poured into a Petri dish and the solvent was evaporated, resulting in a free-standing thin composite film. The XRD analysis of this sample revealed the original d001-spacing of MMT-H. This is an indication, to some degree, of the secondary aggregation of the re-dispersed polar MMT-H platelets from the original fiber exfoliated MMT form. This property is due to the less polar nature of chloroform solvent compared to the protic formic acid solvent. A prominent re-crystallization and increased crystallinity degree of the PHB film matrix after chloroform evaporation was also observed. This is in contrast with the low-crystallinity electrospun PHB-based composite fibers obtained through the fast evaporation of formic acid dispersions explored in this study during the electrospinning process.

In Figure 8, the ATR FT-IR spectra of the prepared electrospun composite membranes are revealed. The band at 1720–1721 cm^{-1} is attributed to the C=O stretch of the ester group. The peak at 1278–1280 cm^{-1} is assigned to the ester bonding observed in the FT-IR spectra of PHB, PHB/MMT 10%, PHB/MMT 20%, and PHB/PVF/MMT 10% [38].

The registered peaks at 2910 cm^{-1} and 1420 cm^{-1} are due to the vibration of C–H from alkyl groups and the bending vibration of CH_2 groups in PVA. The vibration band at 1710 cm^{-1} refers to the stretching C=O vibration from unhydrolyzed acetate groups in the pristine PVA co-polymer. The band at 1089 cm^{-1}, which corresponds to the stretching vibrations of C-O-H in pristine PVA, almost disappeared after in situ esterification with formic acid. The stretching vibration of the carbonyl group associated with the sharp and intensive peak around 1710 cm^{-1} is characteristic for PVF. It shows the high degree of in situ esterification of the PVA precursor. The CH group deformation vibrations of the formyl group are registered at 1380 cm^{-1}. The peak observed at 798 cm^{-1} is attributed to rooking CH_2 vibrations. The band observed at 1167–1174 cm^{-1} is associated with the C-O stretching vibration in PVF, PVF/MMT 30%, and PHB/PVF/MMT 10% [38]. The absorption bands in the range of 515–474 cm^{-1} are due to the deformation vibrations of Si–O-Al and Si–O–Si. In FT-IR spectra of PVF/MMT 30% and PHB/PVF/MMT 10%, an absorption band is observed at 1620 cm^{-1}, associated with the bending vibrations of the -OH group of water molecules present in the montmorillonite [39–41].

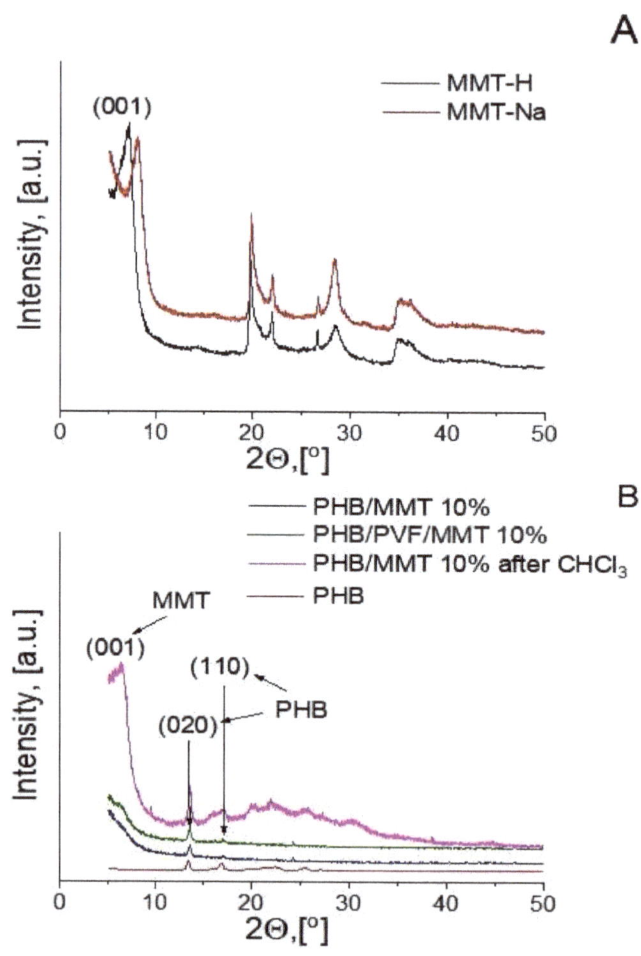

Figure 7. XRD patterns of natural/protonated montmorillonite (**A**) and electrospun composite fibers with 10 wt.% incorporated MMT-H before and the sample PHB/MMT 10% was extracted with chloroform (**B**).

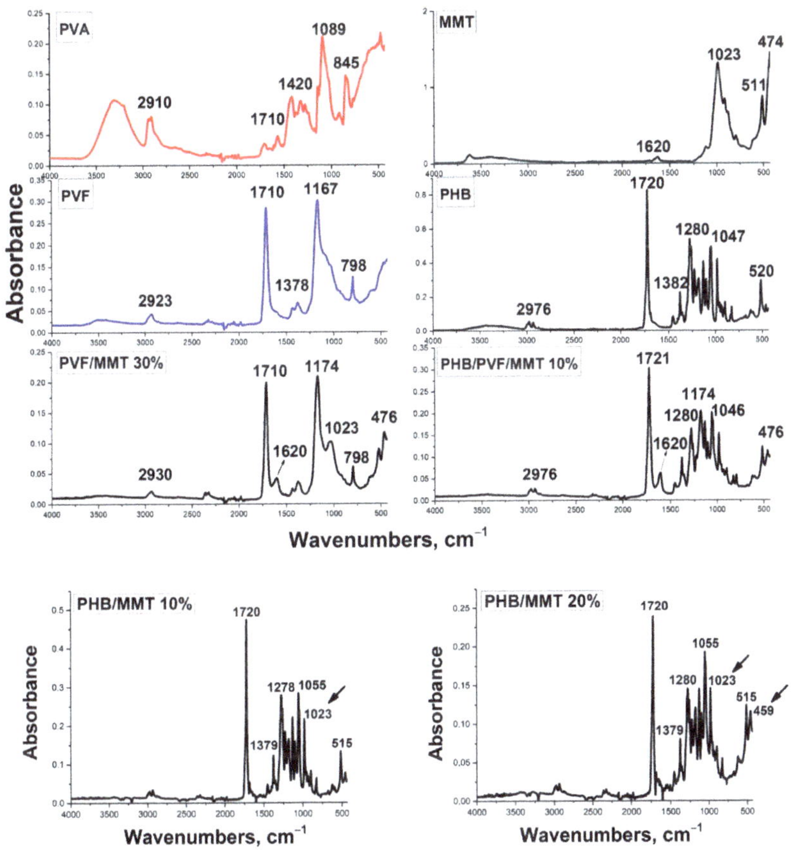

Figure 8. ATR FT-IR analysis of the starting materials and the composite electrospun fibers.

The measured contact angles of the investigated membranes are shown in Figure 9A. Commonly, PVA exhibits a noticeable hydrophilicity. However, after it undergoes an esterification reaction with formic acid, the hydroxyl groups in PVA change into ester groups. Hence, the formed poly(vinyl formate) (PVF) partially loses its hydrophilic behavior, tending to become a hydrophobic matrix. It displays a contact angle of 70.7°. Upon loading PVF with MMT, we can observe a decrease in the contact angle to 57.5°. This lower value signifies a wetting character for this membrane due to the incorporation of hydrophilic MMT. The potent electrostatic impact of the negatively charged surface of MMT allows it to adsorb positively charged surfaces such as cationic dyes. Neat PHB showed a contact angle of 138°. Upon investigating the PHB/MMT 10% sample, it showed a contact angle of 131.4 °C. This value indicates the dominance of a hydrophobic surface owing to PHB in this membrane, even in the presence of MMT. The PHB/PVF/MMT 10% composite had a higher contact angle value of 142.2° when compared to that of PHB/MMT 10%. At higher MMT content, PHB/MMT 20% displayed a CA value of 125°. Figure 9B displays digital images tracking the spread of MB aqueous solution on the surface of the PHB/PVF/MMT 10% composite at different time intervals from t = 0–15 min, showing an adequate wetting behavior due to the presence of MMT. The contact angle value comparison is between the blended PHB/PVF/MMT 10% and PHB/MMT 10% composite mats. In both cases, the predominant hydrophobization role is played by the more hydrophobic PHB homopolymer, rather than the more polar (amphiphilic) PVF ter-polymer, which contains the more polar formate ester, some free -OH and acetate groups. Thus, the observed slight increase in the contact angle/hydrophobicity of the blended PHB/PVF/MMT fiber mats compared to single polymer PHB/MMT 10% could be attributed mostly to the blend fibers' smoother morphology and higher fiber diameter compared with those of PHB/MMT fibers.

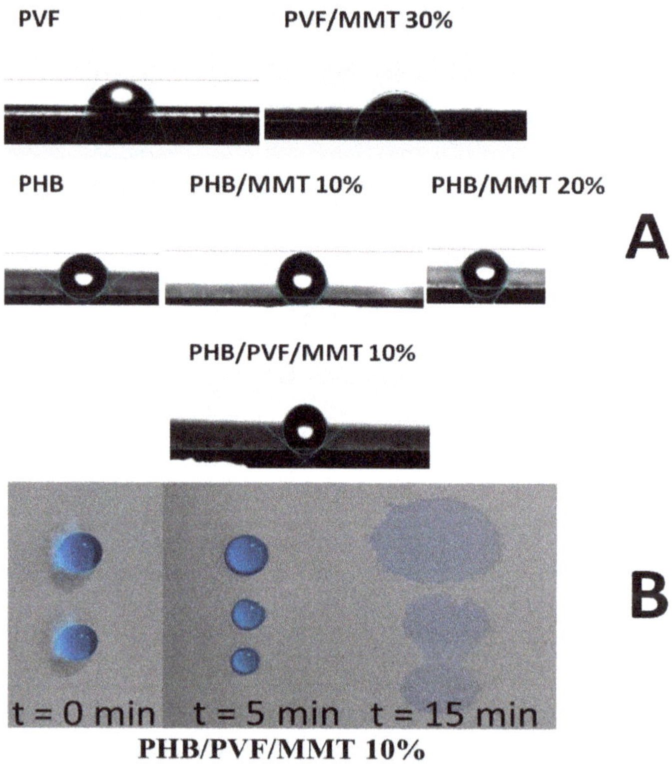

Figure 9. (**A**) Contact angle(s) of PVF, PVF/MMT 30%, neat PHB, PHB/MMT 10%, PHB/MMT 20%, and PHB/PVF/MMT 10%. (**B**) Digital pictures of PHB/PVF/MMT 10% composite membrane in contact with aqueous MB solution at different time intervals.

3.2. Adsorption Isotherms and Kinetics Investigations

All nanofiber mats were evaluated as membranes for removing cationic and anionic dyes (Figure 10). PHB/MMT 20% and PVF/MMT 30% showed significant performance in dye removal compared with the other matrices. PHB/MMT 20% was the best electrospun mat for adsorbing Congo red. Meanwhile, PVF/MMT 30% exhibited the optimum activity for the adsorption of methylene blue and Congo red. This property may be attributed to the presence of two different charges on the mat surface. Moreover, it may be attributed to the high surface area and the highest MMT-H content afforded by the PVF/MMT 30% composite mat, which possesses a smaller nanofiber diameter (average diameter of 157 nm).

Figure 11 indicates the effect of initial dye concentration on the adsorption capacity of the selected nanofiber membranes. The adsorption capacity of methylene blue increased with the elevating the dye concentration, with an enhanced performance for the nanofibers with 30% clay content. Meanwhile, the adsorption capacity boosted upon increasing the Congo red concentration. The adsorption of PVF with 30% clay was higher than the adsorption of fibers with 20% clay, indicating that the higher clay content corresponds to higher adsorption capacity. The increase in dye concentration may be the result of greater diffusion of pollutants into the active site within the adsorbent and increased saturation at these sites. In the case of the dye concentrations of 20 and 30 ppm, the adsorption capacity values were almost close to each other. This behavior may be attributed to the saturation of the monolayer active sites. Then, the adsorption capacity increased by increasing the dye concentration, most likely because of the π-π attraction between the dye molecules. The pollutants may form stalks and make multilayers that lead to an increase in adsorption capacity. Figure 11 also shows that the adsorption of Congo red is much higher than that of methylene blue, pointing to the higher affinity of both polyvinyl alcohol and clay towards anionic dyes. This may be due to the electrostatic attraction between the anionic dyes and the protonated groups of the composites.

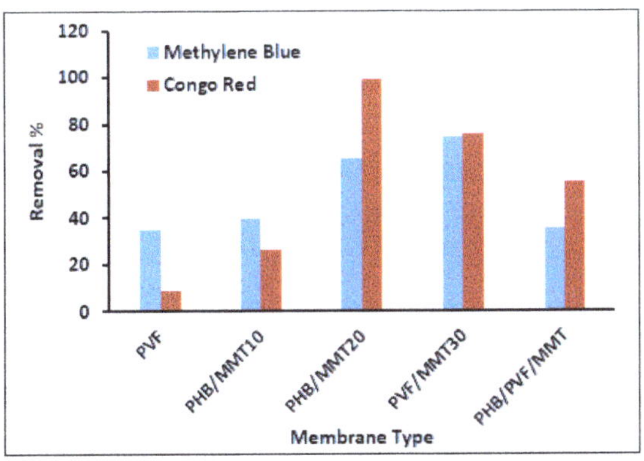

Figure 10. A comparison between the adsorption efficiencies of different electrospun membranes at 25 °C over a period of 24 h with 1 mg of adsorbent and 10 mL of Congo red and methylene blue (10 ppm).

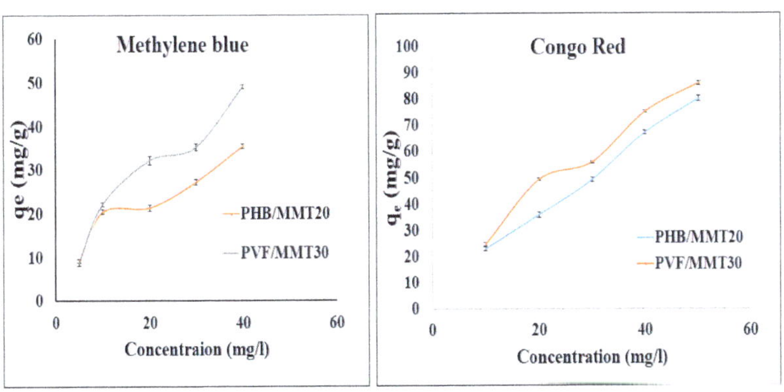

Figure 11. Effect of the concentration of methylene blue and Congo red dyes on adsorption in 1 mg samples of the selected fiber mats, using 10 mL of dye solution at 25 °C over 24 h.

This study examined the interactions between pollutants (dyes) and nanofiber membranes using adsorption isotherm models. The Freundlich and Langmuir models are the two adsorption isotherm models which are commonly used. In order to identify the best model for our purposes, linear regression of the Freundlich and Langmuir models was used. Correlation coefficients (R^2) were estimated and the adsorption isotherm models were evaluated for applicability. Figures 12 and 13 show the linear forms of the Freundlich and Langmuir isotherms for methylene blue and Congo red as models of cationic and anionic dyes, respectively. Generally, the Langmuir isotherm model assumes that monolayer adsorption takes place on surfaces with uniformly distributed adsorption sites and without adsorbate transmigration. Accordingly, once the surface approaches a saturation point, it will achieve maximum adsorption [42,43]. In contrast, the Freundlich model represents an adsorption process on heterogeneous surfaces through multilayer adsorption [44]. Methylene blue adsorption on both membranes is illustrated in Figure 12. The adsorption on the PHB/MMT 20% membrane was fitted with Freundlich and Langmuir isotherms according to correlation coefficient values (R^2) of 0.92 and 0.93, respectively. For PVF/MMT 30%, the Freundlich model has a better fitting (R^2 = 0.98) than the Langmuir model (R^2 = 0.93). As shown in Figure 13, Congo red adsorption on PVF/MMT 30% seems to be fitted with both models. However, there was a higher correlation coefficient value (R^2) for the Freundlich isotherm (0.94) compared to the Langmuir isotherm (0.90). This may suggest that that monolayer adsorption takes place by electrostatic attraction on the active site of PVF/MMT 30%, with multilayers of Congo red stacked over [42]. For PHB/MMT 20%, it is obvious that the adsorption of Congo red is well-fitted to the Freundlich model (R^2 = 0.97). Here, the hydrogen bonding and the π-π bonding between the Congo red molecules assisted in maximizing the adsorption capacity of the PHB/MMT 20% surface.

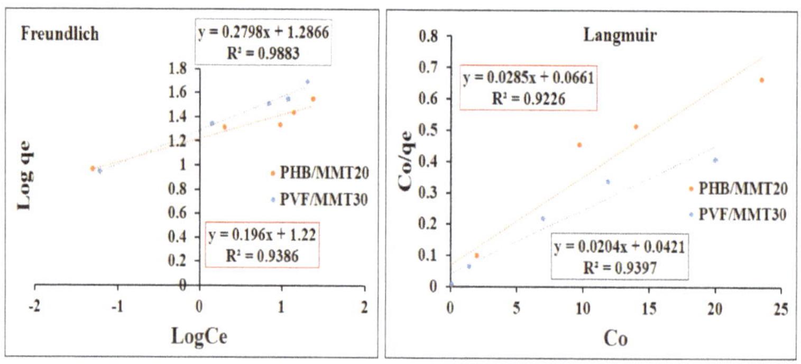

Figure 12. Adsorption Freundlich and Langmuir isotherms for the adsorption of methylene blue on the selected membranes.

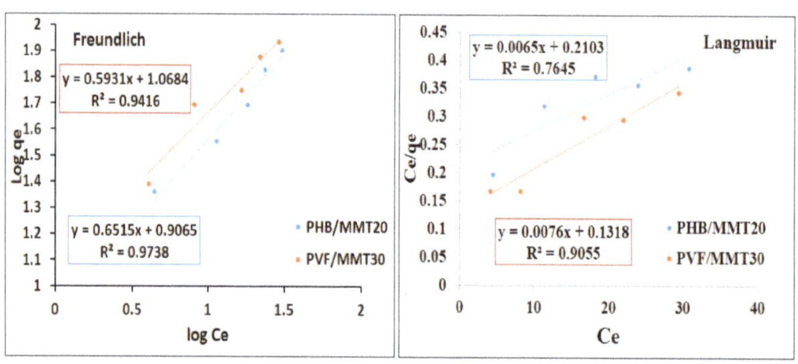

Figure 13. Adsorption Freundlich and Langmuir isotherms for the adsorption of Congo red on the selected membranes.

Figure 14 displays the effect of contact time on the adsorption capacity of the investigated electrospun membranes towards methylene blue and Congo red dyes. It is clearly seen that the maximum adsorption time was attained after 60 min for both methylene blue and Congo red, after which the adsorption capacity reached a steady state. Furthermore, after contact for more than 90 min, the adsorption decreased. This behavior may be correlated to the saturation of the active site of the dye binder. In other words, the site cannot bind anymore, as the dye molecules tend to desorb out from the fiber mat, bringing out a slight decrease in the adsorption capacity. The adsorption capacity likely increased upon elevating the dye concentration. This behavior can be correlated to the chemi- and physi-sorption of dye molecules forming a multilayer of adsorbate. The removal efficiency decreased as the dye concentration increased, with optimum balances at 40 ppm for Congo red and 30 ppm for MB dyes, respectively. Thereafter, the removal percentage decreased extensively and the adsorption capacity decreased slightly. Therefore, this decline in adsorption may be related to the desorption of the aggregated multi-layer dye molecules. After a long time (4 h), the dye molecules adhering through weak Van Der Waals forces started to detach from the matrix.

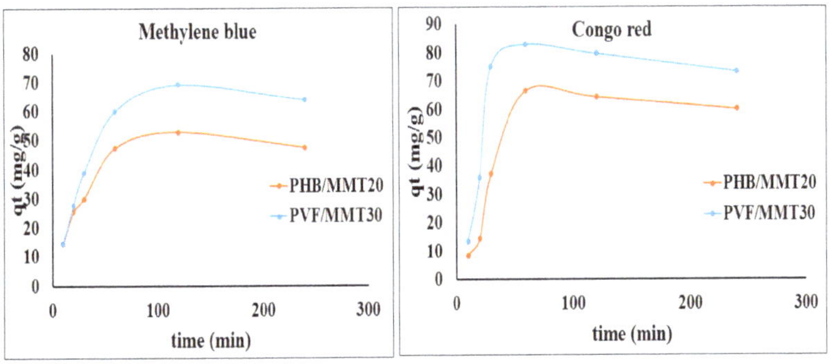

Figure 14. Effect of adsorption time on the adsorption capacity of methylene blue and Congo red using the prepared nanofiber membranes.

Pseudo-first order and pseudo-second order kinetic models were employed to evaluate the dye adsorption rate of the tested membranes. The rates of the adsorption for methylene blue and Congo red on the synthesized membranes were assessed by pseudo-first order and second order kinetics models. The results are shown in Figures 15 and 16. As demonstrated in Figure 15, the adsorption of methylene blue on both PVF/MMT 30% and PHB/MMT 20% fitted with the second order kinetic model, as indicated by the R^2 for both membranes (0.9777 and 0.9836, respectively). The qmax derived from the second order (74 and 52) are closer to the experimental values of 69.1 and 52.7 for the two adsorbents, PVF/MMT 30% and PHB/MMT 20%, respectively. Figure 16 indicates that the rate of adsorption of Congo red was mostly fitted with the second order model for the two membranes, whereas the value of R^2 was 0.9765 and 0.9477 for PVF/MMT 30% and PHB/MMT 20%, respectively. These results indicate that rate of adsorption depends on both the adsorbent nanofibrous membranes and the organic dyes (the cationic methylene blue and the anionic Congo red).

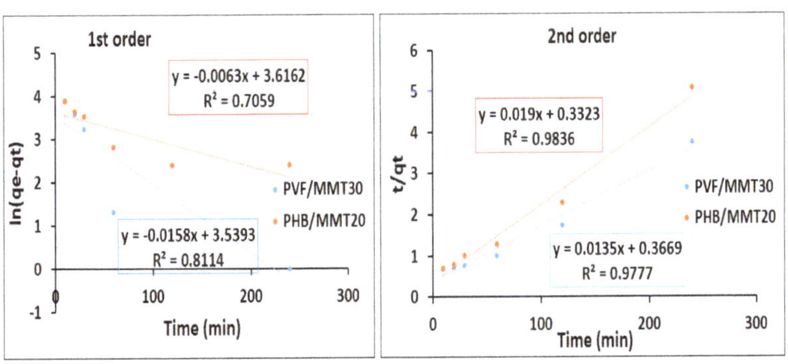

Figure 15. Pseudo-first and second order kinetics for the adsorption of methylene blue using nanocomposite membranes.

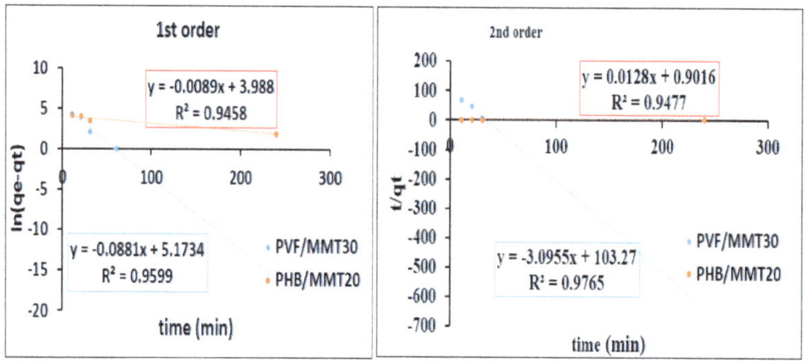

Figure 16. Pseudo-first and second order kinetics for the adsorption of Congo red using nanocomposite membranes.

A proposed adsorption mechanism can be suggested which functions through electrostatic attraction between the active groups of the anionic or cationic dyes with the active sites on the nanofiber surface. Then, the dye molecules accumulate on each other via π-π bonding, as illustrated in Figure 17.

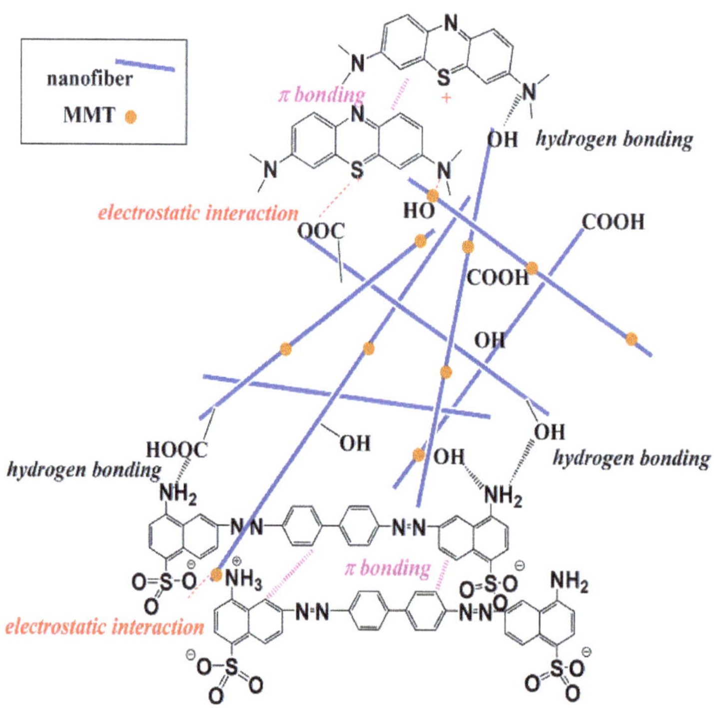

Figure 17. A schematic representation of the proposed adsorption mechanism for the dye on the surface of the nanofiber.

4. Conclusions

In this study, novel composite mats electrospun from PHB and/or in situ PVF, biodegradable polyesters, and their composites, highly loaded with in situ protonated montmorillonite MMT-H, were prepared and characterized. Formic acid was utilized as a reactive solvent for MMT-Na protonation and as an effective clay particle dispersant. The introduction of MMT-H nanofiller into the fibrous membrane polymer matrix leads to the increased hydrophilicity and wettability of the electrospun fibers in contact with aqueous dye media. The dye sorption isotherms and kinetics studies of the investigated nanocomposite fibrous mats showed good potential as biodegradable and disposable fibrous sorbent membranes for the removal from contaminated water of both cationic and anionic dyes, e.g., methylene blue and Congo red. The adsorption capacity of the prepared membranes for MB and CR dye removal boosted noticeably upon raising the MMT content. The adsorption of methylene blue on both PVF/MMT 30% and PHB/MMT 20% fitted with the second order kinetic model, as indicated by R^2 for both membranes (0.9777 and 0.9836, respectively). In addition, the rate of adsorption of Congo red was mostly fitted with the second order model for the two membranes. R^2 showed values of 0.9765 and 0.9477 for PVF/MMT 30% and PHB/MMT 20%, respectively. These findings showed that the adsorption rate relies on both the adsorbent membranes and the organic dyes.

Author Contributions: Conceptualization, H.P., F.U. and A.M.K.; Methodology, H.P. and A.M.K.; Software, K.Z. and D.B.; Validation, D.B., F.U., H.P. and A.M.K.; Formal analysis, H.P., D.B., F.U., G.G. and A.M.K.; Investigation, H.P., D.B., F.U., A.E.A., E.A.A., A.A.E.-S. and A.M.K.; Resources, H.P. and A.M.K.; Data curation, D.B., K.Z., G.G. and F.U.; Writing—original draft preparation, H.P., K.Z., D.B., F.U. and A.M.K.; Writing—review and editing, H.P., A.E.A., E.A.A., A.A.E.-S., N.K., K.Z., F.U., G.G. and A.M.K.; Visualization, K.Z., H.P. and A.M.K.; Supervision, H.P. and A.M.K.; Project administration, H.P. and A.M.K.; Funding acquisition, H.P. and A.M.K. All authors have read and agreed to the published version of the manuscript.

Funding: This research was funded by the Egyptian Academy of Scientific Research & Technology (ASRT) and the Bulgarian Academy of Sciences (BAS). The project's grant numbers are 10115 (Egypt) and IC-EG/05/2022–2024 (Bulgaria).

Data Availability Statement: Data are included in the manuscript.

Acknowledgments: The authors acknowledge the Egyptian Academy of Scientific Research & Technology (ASRT) and the Bulgarian Academy of Sciences (BAS) for funding this project. This work was performed in the frame of joint bilateral project 10115 (Egypt)—IC-EG/05/2022–2024 (Bulgaria). Fiber preparation was partially performed with materials and equipment included in the National Infrastructure NI SEVE supported by the Bulgarian Ministry of Education and Science under grant agreement N° D01-385/D01-18.12.2020. Special thanks to Krasimira Rusinova-Ilieva for the technical support in AFM analysis.

Conflicts of Interest: The authors declare no conflict of interest.

References

1. McAdam, B.; Brennan Fournet, M.; McDonald, P.; Mojicevic, M. Production of polyhydroxybutyrate (PHB) and factors impacting its chemical and mechanical characteristics. *Polymers* **2020**, *12*, 2908. [CrossRef]
2. Vroman, I.; Tighzert, L. Biodegradable Polymers. *Mater. Mol. Divers. Preserv. Int.* **2009**, *2*, 307–344.
3. Utsunomia, C.; Ren, Q.; Zinn, M. Poly(4-Hydroxybutyrate): Current State and Perspectives. *Front. Bioeng. Biotechnol.* **2020**, *8*, 257. [CrossRef]
4. Zheng, Y.; Chen, J.-C.; Ma, Y.; Chen, G.-Q. Engineering Biosynthesis of Polyhydroxyalkanoates (PHA) for Diversity and Cost Reduction. *Metab. Eng.* **2019**, *58*, 82–93. [CrossRef] [PubMed]
5. Avella, M.; Martuscelli, E. Poly-d-(−)(3-hydroxybutyrate)/poly (ethylene oxide) blends: Phase diagram, thermal and crystallization behaviour. *Polymer* **1988**, *29*, 1731–1737. [CrossRef]
6. Ol'khov, A.A.; Iordanskii, A.L.; Shatalova, O.V. Blends based on polyvinyl alcohol and polyhydroxybutyrate. *Int. Polym. Sci. Technol.* **2003**, *30*, 47–50. [CrossRef]
7. Khlyustova, A.; Sirotkin, N.; Kraev, A.; Agafonov, A.; Titov, V. Effect of metal oxides added onto polyvinyl alcohol via pulsed underwater plasma on their thermal, electrical and dielectric properties. *J. Appl. Polym. Sci.* **2021**, *138*, 51174. [CrossRef]
8. Asran, A.S.; Razghandi, K.; Aggarwal, N.; Michler, G.H.; Groth, T. Nanofibers from blends of polyvinyl alcohol and polyhydroxy butyrate as potential scaffold material for tissue engineering of skin. *Biomacromolecules* **2010**, *11*, 3413–3421. [CrossRef]
9. Budurova, D.; Ublekov, F.; Penchev, H. The use of formic acid as a common solvent for electrospinning of hybrid PHB/Soy protein fibers. *Mater. Lett.* **2021**, *301*, 130313. [CrossRef]

10. Herrmann, W.O.; Deisenhofen; Wolfram, H. Process for the Preparation of Polyvinyl Formate. U.S. Patent 2,278,783A, 17 January 1940.
11. Ushakov, S.N.; Kononova, T.A. Synthesis of esters of polyvinyl alcohol. *Bull. Acad. Sci. USSR Div. Chem. Sci.* **1955**, *4*, 103–108 [CrossRef]
12. Essawy, H.A.; Tawfik, M.E.; Khalil, A.M.; El-Sabbagh, S.H. Systematic organophilization of montmorillonite: The impact thereof on the rheometric and mechanical characteristics of NBR and SBR based nanocomposites. *Polym. Eng. Sci.* **2014**, *54*, 942–948. [CrossRef]
13. Bee, S.L.; Abdullah, M.A.A.; Bee, S.T.; Sin, L.T.; Rahmat, A.R. Polymer nanocomposites based on silylated-montmorillonite: A review. *Prog. Polym. Sci.* **2018**, *85*, 57–82. [CrossRef]
14. Chan, M.L.; Lau, K.T.; Wong, T.T.; Ho, M.P.; Hui, D. Mechanism of reinforcement in a nanoclay/polymer composite. *Compos. B Eng.* **2011**, *42*, 1708–1712. [CrossRef]
15. Trigueiro, P.; Pereira, F.A.; Guillermin, D.; Rigaud, B.; Balme, S.; Janot, J.M.; Jaber, M. When anthraquinone dyes meet pillared montmorillonite: Stability or fading upon exposure to light? *Dyes Pigm.* **2018**, *159*, 384–394. [CrossRef]
16. Wang, Y.; Xiong, Y.; Wang, J.; Zhang, X. Ultrasonic-assisted fabrication of montmorillonite-lignin hybrid hydrogel: Highly efficient swelling behaviors and super-sorbent for dye removal from wastewater. *Colloids Surf.* **2017**, *520*, 903–913. [CrossRef]
17. Yotsuji, K.; Tachi, Y.; Sakuma, H.; Kawamura, K. Effect of interlayer cations on montmorillonite swelling: Comparison between molecular dynamic simulations and experiments. *Appl. Clay Sci.* **2021**, *204*, 106034. [CrossRef]
18. Liu, J.; Boo, W.J.; Clearfield, A.; Sue, H.J. Intercalation and exfoliation: A review on morphology of polymer nanocomposites reinforced by inorganic layer structures. *Mater. Manuf.* **2006**, *21*, 143–151. [CrossRef]
19. Sohrabnezhad, S.; Pourahmad, A. Comparison absorption of new methylene blue dye in zeolite and nanocrystal zeolite. *Desalination* **2010**, *256*, 84–89. [CrossRef]
20. Al Kausor, M.; Gupta, S.S.; Bhattacharyya, K.G.; Chakrabortty, D. Montmorillonite and modified montmorillonite as adsorbents for removal of water soluble organic dyes: A review on current status of the art. *Inorg. Chem. Commun.* **2022**, *143*, 109686 [CrossRef]
21. Khalil, A.M.; Kenawy, S.H. Hybrid membranes based on clay-polymer for removing methylene blue from water. *Acta Chim. Slov.* **2020**, *67*, 96–104. [CrossRef]
22. Patel, H.A.; Somani, R.S.; Bajaj, H.C.; Jasra, R.V. Nanoclays for polymer nanocomposites, paints, inks, greases and cosmetics formulations, drug delivery vehicle and waste water treatment. *Bull. Mater. Sci.* **2006**, *29*, 133–145. [CrossRef]
23. Gao, T.; Guan, G.; Wang, X.; Lou, T. Electrospun molecularly imprinted sodium alginate/polyethylene oxide nanofibrous membranes for selective adsorption of methylene blue. *Int. J. Biol. Macromol.* **2022**, *207*, 62–71. [CrossRef]
24. Snoussi, Y.; Sifaoui, I.; El Garah, M.; Khalil, A.M.; Barroso, J.E.; Jouini, M.; Chehimi, M.M. Green, zero-waste pathway to fabricate supported nanocatalysts and anti-kinetoplastid agents from sugarcane bagasse. *Waste Manag.* **2023**, *155*, 179–191. [CrossRef] [PubMed]
25. Almeida, C.A.P.; Debacher, N.A.; Downs, A.J.; Cottet, L.; Mello, C.A.D. Removal of methylene blue from colored effluents by adsorption on montmorillonite clay. *J. Colloid Interface Sci.* **2009**, *332*, 46–53. [CrossRef]
26. Fil, B.A.; Özmetin, C.; Korkmaz, M. Cationic dye (methylene blue) removal from aqueous solution by montmorillonite. *Bull. Korean Chem. Soc.* **2012**, *33*, 10. [CrossRef]
27. Khaniabadi, Y.O.; Basiri, H.; Nourmoradi, H.; Mohammadi, M.J.; Yari, A.R.; Sadeghi, S.; Amrane, A. Adsorption of congo red dye from aqueous solutions by montmorillonite as a low-cost adsorbent. *Int. J. Chem. React. Eng.* **2017**, *16*, 20160203. [CrossRef]
28. Özcan, A.S.; Özcan, A. Adsorption of acid dyes from aqueous solutions onto acid-activated bentonite. *J. Colloid Interface Sci.* **2004**, *276*, 39–46. [CrossRef]
29. Youssef, H.F.; Nasr, R.A.; Abou El-Anwar, E.A.; Mekky, H.S.; Abd El Rahim, S.A. Preparation and characterization of different zeolites from andesite rock: Product evaluation for efficient dye removal. *Microporous Mesoporous Mater.* **2021**, *328*, 111485. [CrossRef]
30. Shahid, M.K.; Kim, Y.; Choi, Y.G. Adsorption of phosphate on magnetite-enriched particles (MEP) separated from the mill scale. *Front. Environ. Sci. Eng.* **2019**, *13*, 1–12. [CrossRef]
31. Shahid, M.K.; Kim, J.Y.; Shin, G.; Choi, Y. Effect of pyrolysis conditions on characteristics and fluoride adsorptive performance of bone char derived from bone residue. *J. Water Process. Eng.* **2020**, *37*, 101499. [CrossRef]
32. Moussout, H.; Ahlafi, H.; Aazza, M.; Maghat, H. Critical of linear and nonlinear equations of pseudo-first order and pseudo-second order kinetic models. *Karbala Int. J. Mod. Sci.* **2018**, *4*, 244–254. [CrossRef]
33. Ublekov, F.; Budurova, D.; Staneva, M.; Natova, M.; Penchev, H. Self-supporting electrospun PHB and PHBV/organoclay nanocomposite fibrous scaffolds. *Mater. Lett.* **2018**, *218*, 353–356. [CrossRef]
34. Olkhov, A.A.; Staroverova, O.V.; Kuherenko, E.L.; Iordanskii, A.L. Effect of electrospinning solution parameters on the properties of nonvolven fibrous material based on polyhydroxibutyrate. *J. Phys. Conf. Ser.* **2020**, *1431*, 012029. [CrossRef]
35. Du, H.; Zhou, T.; Zhang, J.; Liu, X. Moving-window two-dimensional correlation infrared spectroscopy study on structural variations of partially hydrolyzed poly (vinyl alcohol). *Anal. Bioanal. Chem.* **2010**, *397*, 3127–3132. [CrossRef]
36. Sadeghi, M. Synthesis of a biocopolymer carrageenan-g-poly (AAm-co-IA)/montmorilonite superabsorbent hydrogel composite. *Braz. J. Chem. Eng.* **2012**, *29*, 295–305. [CrossRef]

37. Ublekov, F.; Baldrian, J.; Nedkov, E. Crystalline β-structure of PHBV grown epitaxially on silicate layers of MMT. *J. Polym. Sci. B Polym. Phys.* **2009**, *47*, 751–755. [CrossRef]
38. Sindhu, R.; Binod, P.; Pandey, A. Microbial poly-3-hydroxybutyrate and related copolymers. In *Industrial Biorefineries & White Biotechnology*; Elsevier: Amsterdam, The Netherlands, 2015; pp. 575–605.
39. Haken, J.K.; Werner, R.L. The infrared spectrum of polyvinyl formate. *Spectrochim. Acta A Mol. Biomol.* **1971**, *27*, 343–351. [CrossRef]
40. Sharma, P.; Borah, D.J.; Das, P.; Das, M.R. Cationic and anionic dye removal from aqueous solution using montmorillonite clay: Evaluation of adsorption parameters and mechanism. *Desalin. Water Treat.* **2016**, *57*, 8372–8388. [CrossRef]
41. Reddy, C.R.; Nagendrappa, G.; Prakash, B.J. Surface acidity study of Mn+-montmorillonite clay catalysts by FT-IR spectroscopy: Correlation with esterification activity. *Catal. Commun.* **2007**, *8*, 241–246. [CrossRef]
42. Selim, S.E.; Meligi, G.A.; Abdelhamid, A.E.; Mabrouk, M.A.; Hussain, A.I. Novel Composite Films Based on Acrylic Fibers Waste/Nano-chitosan for Congo Red Adsorption. *J. Polym. Environ.* **2022**, *30*, 2642–2657. [CrossRef]
43. Ali, E.A.; Ismail, M.N.; Elsabee, M. Chitosan based polyelectrolyte complexes development for anionic and cationic dyes adsorption. *Egypt. J. Chem.* **2020**, *63*, 537–554. [CrossRef]
44. Abdelhamid, A.E.; Kandil, H. Facile approach to synthesis super-adsorptive hydrogel based on hyperbranched polymer for water remediation from methylene blue. *React. Funct. Polym.* **2022**, *177*, 105312. [CrossRef]

membranes

MDPI

Article

Evaluation of Surface Properties and Separation Performance of NF and RO Membranes for Phthalates Removal

En Qi Lim [1,2], Mei Qun Seah [2], Woei Jye Lau [1,2,*], Hasrinah Hasbullah [1,2], Pei Sean Goh [1,2], Ahmad Fauzi Ismail [1,2] and Daryoush Emadzadeh [3]

[1] Faculty of Chemical and Energy Engineering, Universiti Teknologi Malaysia, Skudai 81310, Johor, Malaysia
[2] Advanced Membrane Technology Research Centre (AMTEC), Universiti Teknologi Malaysia, Skudai 81310, Johor, Malaysia
[3] Department of Chemical and Biological Engineering, University of Ottawa, 161 Louis Pasteur, Ottawa, ON K1N 6N5, Canada
* Correspondence: lwoeijye@utm.my

Abstract: Many studies indicated that phthalates, a common plasticizer, lurk silently in water bodies and can potentially harm living organisms. Therefore, removing phthalates from water sources prior to consumption is crucial. This study aims to evaluate the performance of several commercial nanofiltrations (NF) (i.e., NF3 and Duracid) and reverse osmosis (RO) membranes (i.e., SW30XLE and BW30) in removing phthalates from simulated solutions and further correlate the intrinsic properties of membranes (e.g., surface chemistry, morphology, and hydrophilicity) with the phthalates removal. Two types of phthalates, i.e., dibutyl phthalate (DBP) and butyl benzyl phthalate (BBP), were used in this work, and the effects of pH (ranging from 3 to 10) on the membrane performance were studied. The experimental findings showed that the NF3 membrane could yield the best DBP (92.5–98.8%) and BBP rejection (88.7–91.7%) regardless of pH, and these excellent results are in good agreement with the surface properties of the membrane, i.e., low water contact angle (hydrophilicity) and appropriate pore size. Moreover, the NF3 membrane with a lower polyamide cross-linking degree also exhibited significantly higher water flux compared to the RO membranes. Further investigation indicated that the surface of the NF3 membrane was severely covered by foulants after 4-h filtration of DBP solution compared to the BBP solution. This could be attributed to the high concentration of DBP presented in the feed solution owing to its high-water solubility (13 ppm) compared to BBP (2.69 ppm). Further research is still needed to study the effect of other compounds (e.g., dissolved ions and organic/inorganic matters that might be present in water) on the performance of membranes in removing phthalates.

Keywords: phthalates; commercial membrane; nanofiltration; reverse osmosis; water

1. Introduction

Water pollution caused by plastic has become a serious concern with the development of the plastic industry. In 2018, 359 million tonnes of plastic were produced worldwide, and it is estimated that by 2025, a total of 250 million tonnes of plastics will be discharged [1]. Phthalates have been the well-known plasticizers in the industry since the 1930s due to their high compatibility with polyvinyl chloride (PVC). Apart from being plasticizers, they also act as additives in industrial products such as pesticides, paints, cosmetics, and electrical insulators. Researchers have categorized phthalates as a part of emerging pollutants (EPs) as they can persist for a long time in the environment, accumulate and spread from one species to another via the food chain [2,3].

Phthalates (a.k.a. phthalic acid) are the esters of 1,2-dibenzene dicarboxylic acid derived from a benzene ring and two ester groups [4]. They are transparent or slightly yellowish in appearance and odorless. These features make them difficult to be monitored in the environment, especially water bodies. In general, phthalates show good solubility in most organic solvents but are barely soluble in water. Moreover, they are less volatile and react with other chemical substances. Table 1 shows the physicochemical properties of the six common phthalates, which are di-2-ethylhexyl phthalate (DEHP), diisononyl phthalate (DINP), diisodecyl phthalate (DIDP), di-n-octyl phthalate (DnOP), dibutyl phthalate (DBP), and butyl benzyl phthalate (BBP). Compared to other phthalates, DBP and BBP have a much higher water solubility level and, thus, are selected in this work for investigation. As reported by the United States Environmental Protection Agency (USEPA), phthalates have been found in food and in humans [5]. Several studies reported the case of feminization in male animals and birth defects, especially amphibians, after being exposed to phthalates [6,7]. Additionally, phthalates are one of the endocrine disruptors which bring several negative long-term effects on human reproductive, neurological, and developmental systems [8].

Table 1. Physicochemical properties of the six common phthalates.

Phthalates	Chemical Formula	Molar Mass (g/mol)	Melting Point (°C)	Boiling Point (°C)	Relative Density (Water = 1.0) (g/cm^3)	Solubility in Water (mg/L)	Ref.
Di-2-ethylhexyl phthalate	$C_{24}H_{38}O_4$	390.6	−46	384	0.986	0.270	[9,10]
Diisononyl phthalate	$C_{26}H_{42}O_4$	418.6	−48	252 (at 5 mm Hg)	0.980	0.200	[9,11]
Diisodecyl phthalate	$C_{28}H_{46}O_4$	446.7	−46	253 (at 4 mm Hg)	0.960	0.280	[9,12]
Di-n-octyl phthalate	$C_{24}H_{38}O_4$	390.6	−25	220 (at 2 mm Hg)	0.978	0.022	[9,13]
Dibutyl phthalate (DBP)	$C_{16}H_{22}O_4$	278.4	−35	340	1.050	11.200	[9,14]
Butyl benzyl phthalate (BBP)	$C_{19}H_{20}O_4$	312.4	−35	370	1.100	2.690	[9,15]

Phthalates have undoubtedly brought a lot of convenience to society, but their harmful effects on humans and ecosystems cannot be ignored. A tidal wave of new research has documented their wide-ranging negative health impacts [16]. Emission of phthalates frequently happens as phthalates themselves do not bond covalently to the polymers, thus making them easily enter the environment via water, soil, and air. Phthalates could enter the environment in many ways, such as being lost in the form of solid waste or wastewater in the manufacturing process, diffusing into the atmosphere due to low volatility, and leaching from produced products [4]. Exposure to phthalates easily happens among the population of humans and animals via inhalation, ingestion, and absorption (skin contact).

To minimize exposure rates and meet stricter environmental regulations, several wastewater treatment technologies, including biodegradation, adsorption, and membrane technology, are being considered/used to remove phthalates [17]. Among these methods, membrane technology offers high efficiency for eliminating phthalates with less energy consumption, small space required, and a simple working principle [18]. It can also work together with other technologies, such as adsorption, to enhance its efficiency. There are four types of pressure-driven membranes which are microfiltration (MF, Pore size: 100 nm–2 μm), ultrafiltration (UF, Pore size: 2–100 nm), nanofiltration (NF, Pore size: 1–10 nm) and reverse osmosis (RO, Pore size: <1 nm) [19]. Each of them has different operating pressures, characteristics, and functions.

Currently, there is very little research on using pressure-driven membranes to remove phthalates. Based on the statistics obtained from Scopus, it is shown that there are only about 400 articles mentioning "membranes" and "phthalates" in the article texts over the past five years (years 2018–2022) (Search within Article title, Abstract, Keywords). Owing to the small molecular weight of phthalates (<450 g/mol), most of the researchers used NF or RO membranes to achieve the desired separation rate. In 2004, Bodzek et al. [20] investigated the performance of RO membrane (RO-DS3SE), NF membrane (NF-DS5DK), and UF membrane (UF-DSGM) in removing DEP, DBP, and DEHP from water at different operating pressures (20 bar for RO and NF membranes and 3 bar for UF membrane). Their findings indicated that the NF-DS5CK membrane showed the best phthalates removal performance, recording 99.9% retention for all three phthalates samples [20]. Even though the pore size of the UF membrane is larger than phthalates, this membrane could still record a relatively high removal rate against DEP and DEHP. The authors attributed the results to possible adsorption and/or partial membrane blocking [20].

On the other hand, Ozay et al. [21] conducted an experiment to investigate the efficiency of NF90 and NF90 + BW30 hybrid membrane processes in removing micropollutants and phthalates. The results showed that as high as 99.9% removal could be achieved by the hybrid membrane process (NF90 + BW30) in removing DEP and DEHP. The NF90 membrane, meanwhile, only achieved 97.7% and 98.9% for DEP and DEHP, respectively. Another study found that the NF90 membrane could achieve a higher phthalate (DMP, DEP, and DBP) rejection than NF270 membrane due to its smaller pore sizes [22]. The authors also reported that the fouled membranes tended to increase the phthalate removal efficiency because of the increase in sieving effect and surface hydrophobicity [22].

By taking into consideration the harmful effects of phthalates on humans and ecosystems, the main objective of this work is to evaluate the performance of several commercial membranes (NF and RO) in removing two types of phthalates with high water solubility levels under different testing conditions. All these commercial membranes are composed of PA selective layer supported by a microporous substrate. The cross-linking degree and pore size of the PA layer, however, varies depending on the synthesized conditions used in manufacturing each membrane. We will correlate the surface properties of membranes with the rejection rate of phthalates during the filtration process and investigate how the deposition of phthalates on the membrane surface affects the separation efficiency. Four commercially available membranes are selected (2 NF and 2 RO) for the removal of DBP and BBP at different pH, and the membrane surface characterization is carried out using a series of analytical instruments.

2. Materials and Methods

2.1. Materials

Four commercial thin film composite (TFC) membranes, including two RO (i.e., SW30XLE and BW30 from DuPont FilmTec^{TM}, Wilmington, DE, USA) and two NF membranes (i.e., NF3 from RisingSun Membrane Technology (Beijing) Co., Ltd., China and Duracid from Suez (GE)) were used in this research study. Dibutyl phthalates (DBP) and benzyl butyl phthalates (BBP) with 99% and 98% purity, respectively, were purchased from Sigma-Aldrich, Saint Louis, MI, USA and were used to prepare stock solutions. Sodium sulfate (Na_2SO_4, Merck Sdn Bhd, Selangor, Malaysia) and sodium chloride (NaCl, Merck Sdn Bhd, Selangor, Malaysia) were used for the salt rejection test, while sulfuric acid (H_2SO_4, Merck Sdn Bhd, Selangor, Malaysia) and sodium hydroxide (NaOH, Merck Sdn Bhd, Selangor, Malaysia) were used for solution pH adjustment. All the solutions were prepared by using water (ASTM Type III) produced by a water purification system (Merck Millipore, Burlington, MA, USA).

2.2. Membrane Filtration Performance

Dead-end high-pressure filtration cell (HP4750, Sterlitech, Auburn, WA, USA) made of 316 stainless steel material, as illustrated in Figure 1, was used to carry out the experiments to determine the membrane filtration performances with respect to pure water permeability and salt/phthalates rejection. Prior to the filtration, the respective membrane was cut into circular coupon with an effective surface area of 14.60 cm^2. The membrane coupons were then soaked in pure water and kept at 4 °C until testing. Each membrane was pre-compacted using nitrogen gas for 30 min at a pressure of 11 and 16 bar for NF and RO membranes, respectively, using pure water at a stirring speed of 300 rpm. This compaction is necessary in order to achieve steady-state flux. Then, the pressure was, respectively, reduced to 10 and 15 bar for NF and RO membranes. The measurement was started after 15-min operation at the specific pressure.

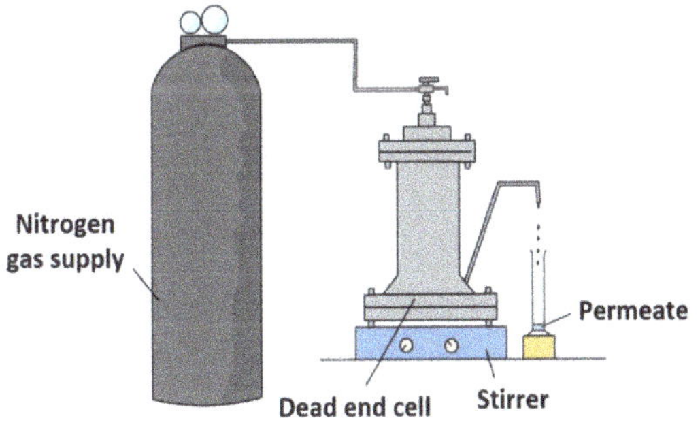

Figure 1. Schematic diagram of dead-end permeation cell setup for membrane evaluation.

The water flux, J_W ($L/m^2{\cdot}h$), and pure water permeability, PWP ($L/m^2{\cdot}h{\cdot}bar$) of each membrane can be evaluated using Equations (1) and (2), respectively.

$$J_W = \frac{V}{A \times t} \tag{1}$$

$$PWP = \frac{J_W}{P} \tag{2}$$

where V is volume of permeate collected (L), A is membrane surface area (m^2), t is taken to collect the permeate (h), and P is operating pressure (bar).

After the PWP test, salt rejection (R_S) of each membrane was investigated by using either 1000-ppm Na_2SO_4 or 1000-ppm NaCl solution. Salt concentration in both feed and permeate was measured by using a benchtop conductivity meter (4520, Jenway, London, UK). Salt rejection, R_S (%), was calculated using Equation (3).

$$R_S = \left| \frac{C_{initial} - C_{final}}{C_{initial}} \right| \times 100 \tag{3}$$

where $C_{initial}$ and C_{final} are the concentration of salt in the feed and permeate, respectively.

The phthalates rejection test was studied using 10 ppm of DBP or 2 ppm of BBP as feed. 5 mL of permeate was collected for both tests, and the time was recorded. The rejection performance of each membrane was examined by using UV-vis spectrophotometer (DR5000, Hach Malaysia Sdn Bhd, Kuala Lumpur, Malaysia). The absorbance of both feed and permeate was measured by scanning them at the peak absorbance wavelength, i.e., 225 nm for DBP and 275 nm for BBP. The rejection of membrane against phthalate, R_P (%), can be determined using the following equation.

$$R_P = \left| \frac{abs_{initial} - abs_{final}}{abs_{initial}} \right| \times 100 \tag{4}$$

where $abs_{initial}$ and abs_{final} are the absorbance of the feed and permeate, respectively.

Single 10 ppm of DBP solution and 2 ppm of BBP solutions were prepared by spiking 10 μL of DBP and 2 μL of BBP into 1000 mL of pure water, respectively. Amount of phthalates added is based on their solubility (DBP: ~13 mg/L and BBP: ~2.69 mg/L) in water. The stock solutions were then sonicated for 7 h in an ultrasonic bath (3800, Branson, Connecticut, USA). To further analyze the performance of the membranes in removing the phthalates, the effect of solution pH was also studied. The pH of the stock solutions (~pH 6) was adjusted to pH 10 and 3 using 1 M of NaOH solution or 0.1 M of H_2SO_4 solution, respectively. The pH value was measured using multiparameter-pH meter (PC2700, Eutech, Singapore). Lastly, the stability and durability of the membrane in removing the phthalates was conducted by running for a duration of 4 h. During the experiment, the flux and rejection were measured and recorded every 30 min.

An adsorption test was also performed to identify the ability of the NF3 membrane to adsorb the phthalates without applying a driving force. In this test, the membrane was placed within a dead-end filtration cell and subject to 2-h stirring at 500 rpm without having external driving force. The absorbance of the single DBP and BBP solutions in the cell before and after adsorption test was then measured to calculate the reduction of phthalates due to membrane adsorption.

2.3. Membrane Characterization

Attenuated total-reflectance Fourier transform infrared spectroscopy (ATIR-FTIR) (Nicolet iS10, Thermo Fisher Scientific, Waltham, MA, USA) was applied to identify the functional group of the membranes at wavenumber ranging 500–4000 cm^{-1}. An average of 32 scans was performed to yield the spectrum of each membrane. Surface roughness of the dried membranes was determined using atomic force microscopy (AFM) (NX10, Park Systems, Suwon, Republic of Korea). The hydrophilicity of membranes was evaluated via the sessile drop method using contact angle goniometer (OCA15Pro, Data Physics, San Jose, CA, USA). At least 10 measurements were performed on each membrane sample in order to yield the average value. A water droplet (pure water) of 0.5 μL was placed on the dried membrane surface for each measurement. To visualize the change of membrane surface before and after the filtration process, scanning electron microscopy (SEM) (S-3400N, Hitachi, Tokyo, Japan) was used. Prior to SEM analysis, the samples were sputter-coated with a thin layer of gold so better images could be produced.

3. Results and Discussion

3.1. Surface Chemistry and Surface Morphology of Membranes

Figure 2a,b shows the FTIR spectra of the commercial RO and NF membranes used in this work, respectively. The results indicated that the commercial TFC RO and NF membranes used were composed of a polyamide (PA) layer due to the presence of characteristics peaks of the PA layer, including C-N stretching (1583 and 1584 cm^{-1}) and amide I and C=O stretching (1647, 1659, 1660, and 1662 cm^{-1}). Comparing between two RO membranes, it was found that the BW30 membrane exhibited higher intensity than the SW30XLE membrane for the broad peak that appeared in the region of 3000–3500 cm^{-1}, and this was mainly due to its lower PA crosslinking degree, which produced higher amount of –OH groups [23]. For the NF membranes, it was found that both NF3 and Duracid membranes displayed very similar peaks, except an additional peak at 1737 cm^{-1} was detected in the NF3 membrane. The additional peak could be due to the presence of carbonyl groups (C=O) resulting from the use of different additives during PA layer synthesis. It must be pointed out that the exact synthesis conditions are remained largely unknown to the public mainly due to the trade secret in manufacturing the membrane products.

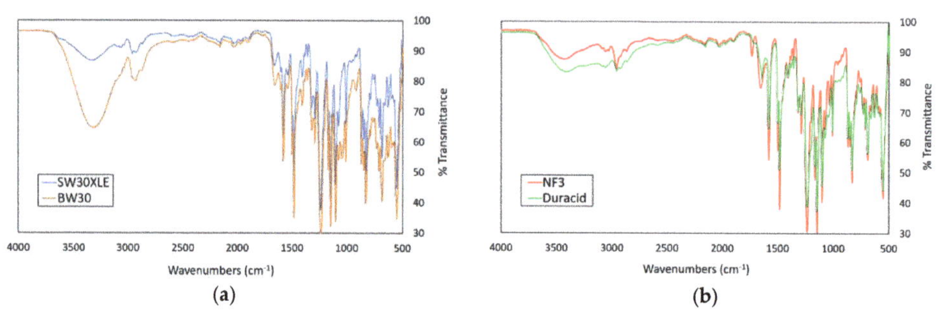

Figure 2. The ATR-FTIR spectra of different membranes, (**a**) SW30XLE and BW30 (RO type) and (**b**) NF3 and Duracid (NF type).

Figure 3 presents the 3D and 2D images of the selective layer of the membranes studied. Generally, all the membranes exhibited a relatively smooth surface with a R_a value < 10 nm, except for the SW30XLE membrane with a significantly rougher surface (R_a: 68.86 nm). Our findings agreed with the work of Ebrahim et al. [24], in which the researchers also reported a very similar roughness value of the SW30XLE membrane, i.e., 78 nm. From the figure, the SW30XLE and Duracid membranes display typical nodular morphologies with obvious hills and valleys. However, the Duracid membrane has a significantly lower R_a value than the SW30XLE membrane. The roughness value of the BW30 membrane was similar to the Duracid membrane, even though their surface morphology was completely different. Among the membranes studied, the NF3 membrane displayed the lowest surface roughness, i.e., 2.45 nm. It must be noted that surface roughness will not only affect the membrane water flux but also govern the solutes deposition/adsorption on the membrane surface. The rougher surface was reported to have a higher fouling tendency as it provides larger contact areas for solutes to deposit [25,26].

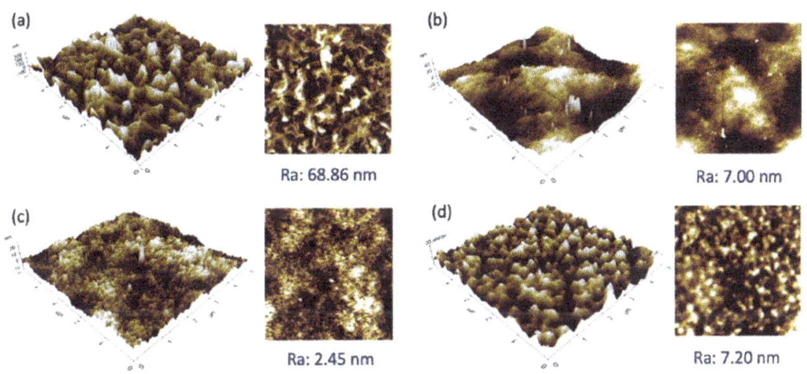

Figure 3. 3D (left) and 2D (right) AFM images of RO and NF membranes together with R_a value, (**a**) SW30XLE, (**b**) BW30, (**c**) NF3, and (**d**) Duracid.

3.2. Pure Water Permeability, Contact Angle, and Salt Rejection of Membranes

The PWP and water contact angle of four different types of commercial TFC membranes are presented in Figure 4a. As can be seen, the water contact angle on the membrane is in the order of BW30 (74.46°) > Duracid (67.97°) > SW30XLE (31.50°) > NF3 (25.72°). The water contact angle is correlated with the hydrophilicity of the TFC membrane, where the smaller the contact angle, the higher the surface hydrophilicity. A membrane with greater surface hydrophilicity will allow more water to pass through its matrix and thus results in higher water flux. However, the pore size of the membrane will also play a role in determining the membrane water flux in addition to surface hydrophilicity. For instance, the water contact angle of SW30XLE is the second lowest among the membrane studies, but such good hydrophilicity does not really contribute to high water permeance as its dense structure is the dominant factor governing water transport. Our findings indicated that all the TFC membranes exhibit hydrophilic characteristics as their water contact angles are <80°.

With respect to PWP, it is obvious that the NF3 membrane recorded the highest value (12.69 L/m^2.h.bar). Its promising water permeance could be attributed to its lowest water contact angle (compared to other membranes) and larger surface pore size than the RO-type membranes [27]. Due to its high PWP, the NF3 membrane requires less operational energy and system footprint to achieve the same water productivity as other membranes. Additionally, its greater surface hydrophilicity could play a role in minimizing the foulants deposition on its surface [28,29]. The lowest PWP of the Duracid membrane could be attributed to its denser selective layer due to the addition of sulfonyl chloride ($-SO_2Cl$) on the PIP-TMC-based membrane [29,30]. Furthermore, the support layer of the Duracid membrane was reported to be significantly thicker compared to other NF membranes (DK, DL, and NF270) tested in the work of Gao et al. [29]. Its support layer is around two times thicker than the support of typical NF membranes. This thick structure tends to create greater resistance for water to transport and thus leads to lower water flux.

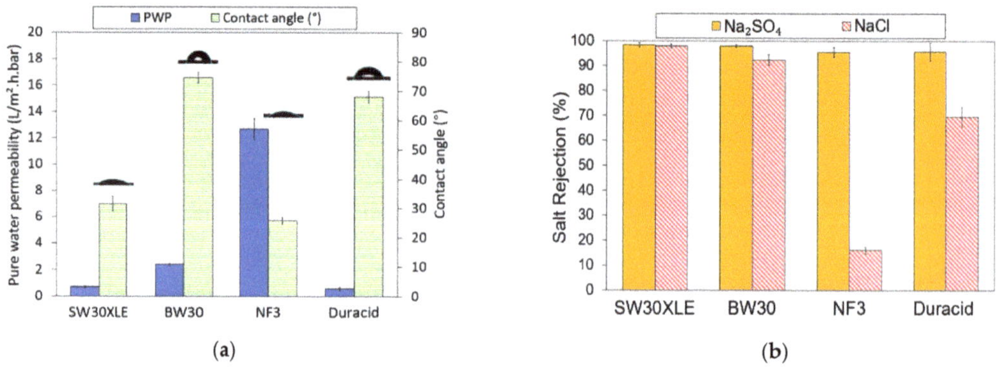

Figure 4. Properties of RO and NF membrane, (**a**) Pure water permeability (PWP) and water contact angle and (**b**) Monovalent and divalent salt rejection.

With respect to the salt rejection, as shown in Figure 4b, it can be observed that all the TFC membranes exhibited promising Na_2SO_4 rejection (95.5–98.3%). In comparison, both RO membranes (SW30XLE and BW30) performed better than the NF membranes (NF3 and Duracid) in removing Na_2SO_4, recording at least 97.9% rejection. Moreover, the SW30XLE and BW30 membranes also achieved significantly higher NaCl rejection of 97.9% and 92.3%, respectively. This is because the RO membranes are designed for the purposes of brackish/seawater desalination process and aim to purify the water by rejecting the passage of monovalent salt. Compared to the BW30 membrane, the SW30XLE membrane demonstrated better performance in removing dissolved ions owing to its higher degree of PA cross-linking, which formed a denser skin layer. On the other hand, the NF membranes showed lower salt rejection rates compared to the RO membrane. This was mainly due to the larger surface pore size, which was not effective in retaining a small hydrated radius of Na and Cl ions [31].

3.3. *Membrane Performance for Phthalates Removal*

3.3.1. Membrane Rejection against DBP and BBP

Figure 5 shows the performance of commercial TFC membranes for the removal of two different types of phthalates. The experiment was separately conducted using DBP and BBP solution with pH at ~6. Experimental findings showed that both SW30XLE and NF3 membranes were able to achieve a relatively constant rejection against both DBP and BBP. However, upon comparing to the NF3 membrane, the SW30XLE membrane demonstrated a significantly higher rejection against DBP (98.5%) and BBP (100%). The excellent performance could be attributed to the high cross-linking degree of the polyamide layer.

The SW30XLE membrane Is a typical m-phenylene diamine–trimesoylchloride (MPD-TMC) membrane with a full aromatic polyamide structure due to the absence of a dominant peak at 1730 cm^{-1} [32]. This selective layer offers promising rejection against small solutes, including monovalent ions, while allowing water to flow through [33]. Higher PA crosslinking reduces the gap space between polymer chain segments so that the segment vibration is limited, thereby improving selectivity [34]. Additionally, a highly cross-linked membrane tends to have a reduced swelling effect under neutral conditions due to high mechanical strength [34]. Hence, its structure, especially the pore size, will not be easily altered.

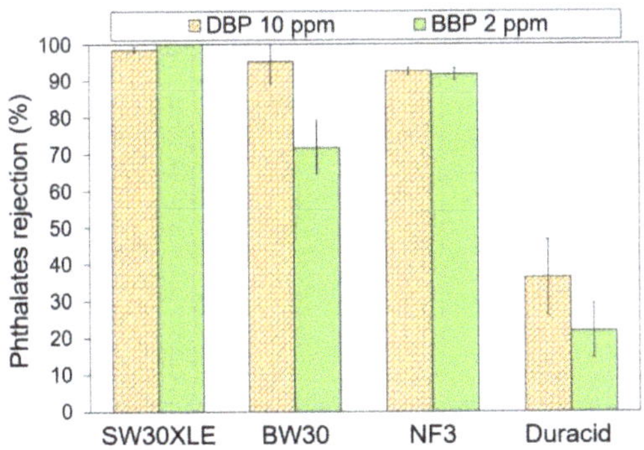

Figure 5. Phthalates rejection by different RO and NF membranes at ~pH 6.30.

On the other hand, the NF3 membrane is a piperazine–trimesoylchloride (PIP-TMC) based membrane with relatively good surface hydrophilicity. Its water contact angle is the lowest one compared to the rest of the membranes tested in this work. Since the log K_{ow} of both phthalates are between 4.50 and 4.73, these chemicals are typically considered semi-hydrophobic. The hydrophilic (membrane)-hydrophobic (phthalates) interaction, in addition to the small pore size of the membrane, could create repulsion, which is beneficial to reject phthalates. Nevertheless, the rejection recorded by the NF3 membrane is lower than the SW30XLE membrane, and this is mainly due to its larger surface pore size, which is mainly used to separate divalent salt instead of monovalent salt [35]. With respect to membrane surface charge, this factor might not play a significant role in influencing phthalate removal as both DBP and BBP do not really carry charge at neutral conditions.

Compared to the SW30XLE and NF3 membranes, the performance of the BW30 membrane in rejecting phthalates is inconsistent as it records a high DBP rejection (95.3%) but a low BBP rejection (71.8%). The reduced separation efficiency of the BW30 membrane compared to the SW30XLE membrane is due to its looser polyamide layer, as the membrane is designed to desalinate brackish water instead of seawater. Furthermore, our findings showed that the Duracid was not suitable for treating phthalates-contaminated solutions owing to its extremely low phthalates separation rate, recording 36.4% and 21.9% for DBP and BBP, respectively. Based on the membrane specifications from the manufacturer [36], the Duracid membrane is suitable to be applied to purify acid-containing wastewater and concentrate metals at low pH streams instead of removing organic pollutants in a neutral environment.

3.3.2. Effect of pH on DBP and BBP Removal

Industrial wastewater typically has varying pH depending on the types of contaminants, chemicals, and microorganisms present in the solution. Therefore, it is also important to study to what extent the solution pH could affect the membrane performance in removing phthalates. Figure 6 presents the membrane performance in rejecting DBP and BBP at different pHs. In this section, only three membranes (i.e., SW30XLE, BW30, and NF3) were studied. We excluded the Duracid membrane from the investigation because of its extremely low phthalates rejection, as reported in the previous section.

As shown in Figure 6, the phthalates rejection performance of the NF3 membrane was relatively higher than the SW30XLE and BW30 membrane at pH 3 and 10. Additionally, it was observed that the NF3 membrane also exhibited more consistent phthalates removal at a wide-range pH condition than other membranes. A study revealed that the NF3 membrane (made of PIP-TMC) exhibited high chemical stability at pH between 2 and 11 [37]. Our experimental findings indicated that the NF3 membrane was able to remove >90% of DBP and 88% of BBP at a pH ranging from 3 to 10. Conversely, the performances of both SW30XLE and BW30 membranes were negatively affected when they were used in acid (pH 3) and alkali (pH 10) conditions. It is believed that the NF3 membrane, which is made of PIP-TMC, could be more suitable than the MPD-TMC-based RO membranes for phthalates removal owing to the better hydrophilic (membrane)-hydrophobic (phthalates) interaction resulted from its lowest water contact angle.

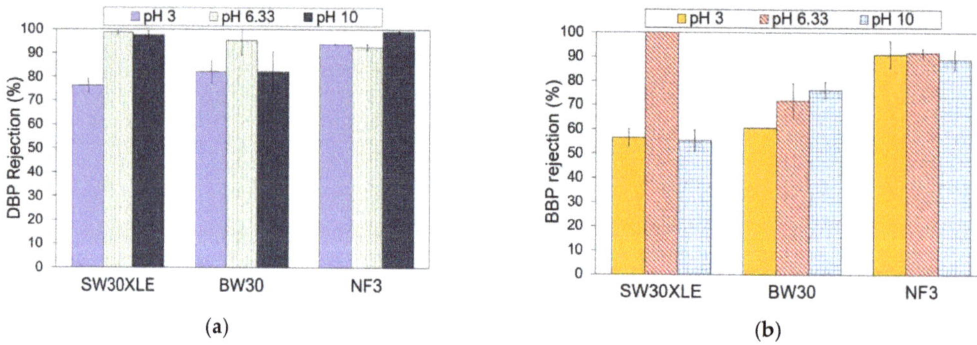

Figure 6. Effect of pH on the membrane performance for the removal of (**a**) DBP and (**b**) BBP.

In general, the DBP rejection was adversely affected when the feed solution was acidic (pH 3). A low pH condition could hydrolyze DBP into protonated phthalic acid, which increases the number of proton charges in DBP [38]. Because of this, the rejection of the membrane against DBP tends to decrease owing to the increase in attractive force between the positively charged DBP and the negatively charged membranes, increasing the passage of DBP through the membrane.

3.3.3. Quality of Treated Water

Figure 7 compares the concentration of DBP and BBP in the permeate produced by the NF3 membrane at different pH. A significant decrease in the DBP concentration in the feed solution was observed, where the concentration was reduced from 10 ppm to 0.753 ppm at ~pH 6.33. While under extreme acid and alkali conditions, the NF3 membrane was also able to achieve DBP concentrations of 0.625 ppm and 0.116 ppm, respectively. With respect to the BBP, the NF3 membrane could attain a very low concentration of the phthalate, recording at 0.179–0.222 ppm. The detection of lower BBP concentration compared to the DBP concentration was mainly due to the low feed concentration of BBP used as a result of its low water solubility. It was found that the NF3 membrane was capable of producing a permeate with DBP and BBP concentrations lower than the threshold limit value (5 ppm) set by the United States Environmental Protection Agency (USEPA) for phthalates [39].

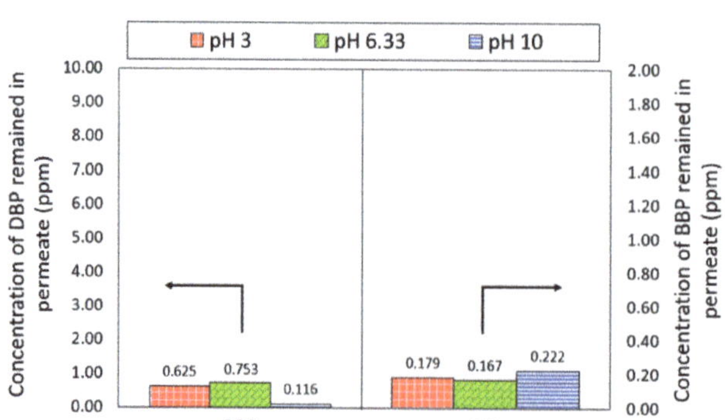

Figure 7. Concentration of DBP and BBP in the permeate of NF3 membrane.

3.4. Short-term Performance Stability of NF3 Membrane

The performance of the NF3 membrane against DBP and BBP removal was further evaluated as a function of filtration time, and the results are presented in Figure 8. As shown, the rejection of the NF3 membrane tended to decrease as a function of time. Its DBP rejection decreased from almost 100% at the early of the filtration process to 83.1% at the end of the experiment. In comparison, the rejection of membrane against BBP also decreased, but its performance was approximately stable after a 120-min operation. With respect to water flux, the membrane displayed decreasing flux in filtering both DBP and BBP. However, the membrane suffered from more severe flux deterioration in filtering DBP compared to the filtration of BBP. The decrease in membrane water flux could be due to surface fouling, while the decrease in phthalates could be partly due to the adsorption behavior of the membrane. It is believed that besides the sieving and charge effects, the adsorption of phthalates onto the membrane itself was also contributing to the separation. Once the membrane's adsorption against phthalates is near saturation, more phthalates are likely to permeate through the membrane, which leads to lower rejection. Based on the findings obtained from 2-h adsorption, it was found that the NF3 membrane could reduce the phthalate concentration in water by 17.74% and 47.3% for DBP and BBP, respectively.

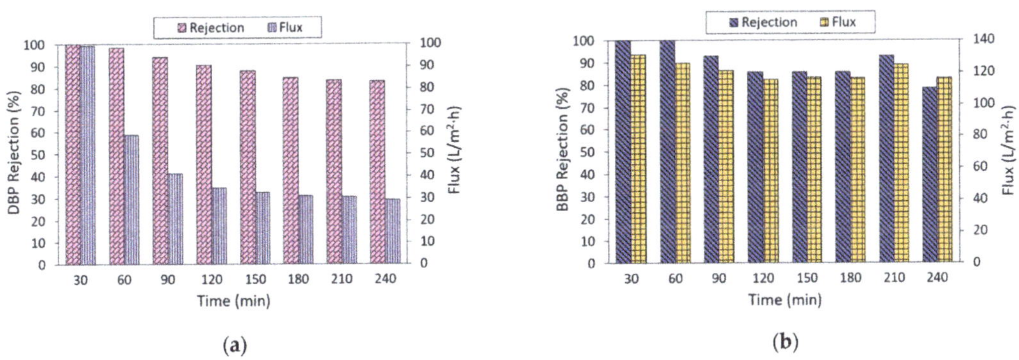

Figure 8. Performance stability of NF3 membrane as a function of time in filtrating (**a**) DBP (10 ppm) and (**b**) BBP (2 ppm) solution at pH ~6.3.

Based on the FTIR spectra shown in Figure 9a, it was obvious that several peaks of the pristine NF3 membrane were altered after it was used for DBP and BBP filtration. The intensities of peaks at 3000–3500 cm^{-1} region and ~1737 cm^{-1} were reduced, and this was mainly due to the membrane surface fouling caused by the deposition/adsorption of phthalates. The SEM images in Figure 9b indicated that the surface of the used membranes (for DBP and BBP removal) was deposited with particles, and significantly large particles were detected on the surface of the membrane used for DBP removal. The findings were in good agreement with the significant flux decline in the membrane used for DBP removal compared to the one used for BBP removal. Even though the feed concentration of phthalates was low for the experiment, the low solubility of DBP and BBP in water could easily cause them to form particles, especially on the membrane surface. This can be explained by the fact that when the membrane rejects phthalates during the filtration process, the accumulation of phthalates on the membrane surface could lead to a significant increase in phthalates concentration, exceeding its solubility. Because of this reason, phthalates can form particles on the membrane surface and causes fouling.

Comparing the SEM surface image of the membrane used to treat DBP-containing water, the membrane used for BBP filtration did not have much fouling on its surface. Less fouling caused by BBP could be due to its lower water solubility (2.69 ppm) than that of DBP (13 ppm). In this study, DBP and BBP solutions were prepared by spiking 10 µL of pure DBP and 2 µL of pure BBP chemicals into 1-L RO water, respectively. Obviously, the amount of BBP particles that are left in the filtration cell would certainly be fewer than DBP particles during the membrane filtration process.

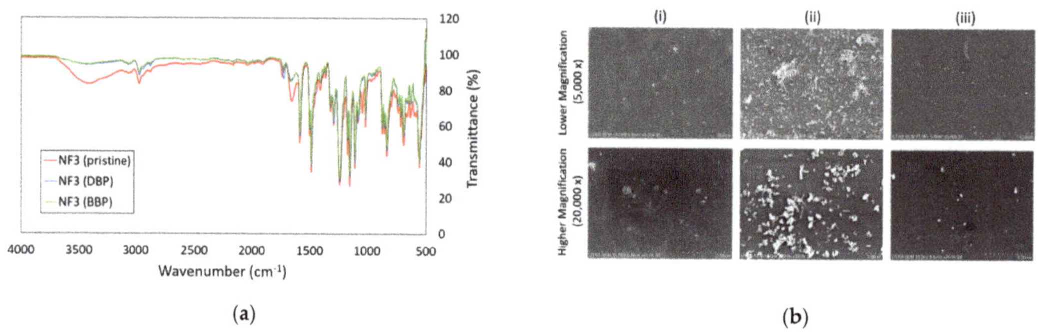

Figure 9. (**a**) FTIR spectrum of pristine NF3 membrane before and after DBP and BBP filtration tests and (**b**) SEM surface images of (i) Pristine NF3 membrane, (ii) NF3 membrane used for DBP filtration and (iii) NF3 membrane used for BBP filtration at two magnifications (5000× and 20,000×).

4. Conclusions

In this work, we evaluated the performance of four commercial TFC membranes in removing two selected phthalates from an aqueous solution and correlated the surface properties of the membrane with removal efficiency. Although all the membranes were made of a polyamide layer supported by a substrate, the membrane surface properties varied depending on the synthesis conditions of each manufacturer. Among the four membranes tested, our results showed that the NF3 membrane was the most promising membrane by taking into account its high DBP (92.5–98.8%) and BBP rejection (88.7–91.7%) regardless of solution pH. This membrane could also purify phthalates-contaminated water with a wide pH range (3–10) to a safe level as stipulated in the USEPA regulations (<5 ppm). Furthermore, the NF3 membrane also exhibited the highest water permeance compared to the rest of the membranes. The excellent filtration characteristics of the NF3 membrane could be attributed to its good surface hydrophilicity coupled with a looser skin layer and its negative charge feature. However, further investigation indicated that the surface of the NF3 membrane was severely covered by foulants after 4-h filtration of DBP solution compared to the BBP solution. This could be attributed to the high concentration of DBP presented in the feed solution owing to its high-water solubility (13 ppm) compared to BBP (2.69 ppm). It must also be pointed out that further research is still needed to study the effect of other compounds (e.g., dissolved ions, organic/inorganic matters, etc., that might be present in water) on the performance of membranes in removing phthalates.

Author Contributions: E.Q.L.: methodology, investigation, formal analysis, data curation, writing—original draft preparation. M.Q.S.: conceptualization, methodology, investigation. W.J.L.: conceptualization, writing—review and editing, supervision, resources. H.H.: writing—review and editing, validation. P.S.G.: writing—review and editing, validation. A.F.I.: writing—review and editing, validation, resources. D.E.: writing—review and editing, validation. All authors have read and agreed to the published version of the manuscript.

Funding: This research received no external funding.

Institutional Review Board Statement: Not applicable.

Informed Consent Statement: Not applicable.

Data Availability Statement: Data sharing is not applicable.

Conflicts of Interest: The authors declare no conflict of interest.

References

1. Becky Miriyam, I.; Anbalagan, K.; Magesh Kumar, M. Phthalates Removal from Wastewater by Different Methods—A Review. *Water Sci. Technol.* **2022**, *85*, 2581–2600. [CrossRef] [PubMed]
2. Dolatabadi, M.; Ahmadzadeh, S. Phthalates as Emerging Pollutants in Water Environment: Control & Treatment Strategies. *J. Environ. Health Sustain. Dev.* **2018**, *3*, 554–556.
3. Abtahi, M.; Dobaradaran, S.; Torabbeigi, M.; Jorfi, S.; Gholamnia, R.; Koolivand, A.; Darabi, H.; Kavousi, A.; Saeedi, R. Health Risk of Phthalates in Water Environment: Occurrence in Water Resources, Bottled Water, and Tap Water, and Burden of Disease from Exposure through Drinking Water in Tehran, Iran. *Environ. Res.* **2019**, *173*, 469–479. [CrossRef]
4. Peijnenburg, W.J.G.M. Phthalates. In *Encyclopedia of Ecology, Five-Volume Set*; Academic Press: Cambridge, MA, USA, 2008; pp. 2733–2738, ISBN 9780080914565.
5. US EPA Risk Management for Phthalates. Available online: https://www.epa.gov/assessing-and-managing-chemicals-under-tsca/risk-management-phthalates (accessed on 14 January 2023).
6. Park, C.B.; Kim, G.E.; Kim, Y.J.; On, J.; Park, C.G.; Kwon, Y.S.; Pyo, H.; Yeom, D.H.; Cho, S.H. Reproductive Dysfunction Linked to Alteration of Endocrine Activities in Zebrafish Exposed to Mono-(2-Ethylhexyl) Phthalate (MEHP). *Environ. Pollut.* **2020**, *265*, 114362. [CrossRef] [PubMed]
7. Bissegger, S.; Pineda Castro, M.A.; Yargeau, V.; Langlois, V.S. Phthalates Modulate Steroid 5-Reductase Transcripts in the Western Clawed Frog Embryo. *Comp. Biochem. Physiol. Part C Toxicol. Pharmacol.* **2018**, *213*, 39–46. [CrossRef]
8. Wang, Y.; Qian, H. Phthalates and Their Impacts on Human Health. *Healthcare* **2021**, *9*, 603. [CrossRef]
9. Raouf, M.E.A.; Maysour, N.E.; Farag, R.K.; Abdul-Raheim, A.M. Wastewater Treatment Methodologies, Review Article. *Int. J. Environ. Agric. Sci.* **2019**, *3*, 18.
10. National Center for Biotechnology Information PubChem Compound Summary for CID 8343, Bis(2-Ethylhexyl) Phthalate. Available online: https://pubchem.ncbi.nlm.nih.gov/compound/Bis_2-ethylhexyl_-phthalate (accessed on 25 February 2023).
11. National Center for Biotechnology Information PubChem Compound Summary for CID 590836, Diisononyl Phthalate. Available online: https://pubchem.ncbi.nlm.nih.gov/compound/Diisononyl-phthalate (accessed on 25 February 2023).
12. National Center for Biotechnology Information PubChem Compound Summary for CID 33599, Diisodecyl Phthalate. Available online: https://pubchem.ncbi.nlm.nih.gov/compound/Diisodecyl-phthalate (accessed on 25 February 2023).
13. National Center for Biotechnology Information PubChem Compound Summary for CID 8346, Dioctyl Phthalate. Available online: https://pubchem.ncbi.nlm.nih.gov/compound/Dioctyl-phthalate (accessed on 25 February 2023).
14. National Center for Biotechnology Information PubChem Compound Summary for CID 3026, Dibutyl Phthalate. Available online: https://pubchem.ncbi.nlm.nih.gov/compound/Dibutyl-phthalate (accessed on 25 February 2023).
15. National Center for Biotechnology Information PubChem Compound Summary for CID 2347, Benzyl Butyl Phthalate. Available online: https://pubchem.ncbi.nlm.nih.gov/compound/Benzyl-butyl-phthalate (accessed on 25 March 2023).
16. Westervelt, A. Phtalates Are Everywhere, and the Health Risks Are Worrying. How Bad Are They Really? *Guardian* **2015**.
17. Vieira, W.T.; de Farias, M.B.; Spaolonzi, M.P.; da Silva, M.G.C.; Vieira, M.G.A. Removal of Endocrine Disruptors in Waters by Adsorption, Membrane Filtration and Biodegradation. A Review. *Environ. Chem. Lett.* **2020**, *18*, 1113–1143. [CrossRef]
18. Algieri, C.; Pugliese, V.; Coppola, G.; Curcio, S.; Calabro, V.; Chakraborty, S. Arsenic Removal from Groundwater by Membrane Technology: Advantages, Disadvantages, and Effect on Human Health. *Groundw. Sustain. Dev.* **2022**, *19*, 100815. [CrossRef]
19. Khulbe, K.C.; Feng, C.Y.; Matsuura, T. Pore Size, Pore Size Distribution, and Roughness at the Membrane Surface. In *Synthetic Polymeric Membranes: Characterization by Atomic Force Microscopy*; Springer: Berlin/Heidelberg, Germany, 2008; pp. 101–139, ISBN 978-3-540-73994-4.
20. Bodzek, M.; Dudziak, M.; Luks-Betlej, K. Application of Membrane Techniques to Water Purification. Removal of Phthalates. *Desalination* **2004**, *162*, 121–128. [CrossRef]
21. Ozay, Y.; Canli, O.; Unal, B.O.; Keskinler, B.; Dizge, N. Investigation of Plasticizer Production Industry Wastewater Treatability Using Pressure-Driven Membrane Process. *Water Supply* **2021**, *21*, 1994–2007. [CrossRef]
22. Zhu, L. Rejection of Organic Micropollutants by Clean and Fouled Nanofiltration Membranes. *J. Chem.* **2015**, *2015*, 934318. [CrossRef]
23. Ng, Z.C.; Lau, W.J.; Ismail, A.F. GO/PVA-Integrated TFN RO Membrane: Exploring the Effect of Orientation Switching between PA and GO/PVA and Evaluating the GO Loading Impact. *Desalination* **2020**, *496*, 114538. [CrossRef]
24. Abdulsalam Ebrahim, M.; Karan, S.; Livingston, A.G. On the Influence of Salt Concentration on the Transport Properties of Reverse Osmosis Membranes in High Pressure and High Recovery Desalination. *J. Membr. Sci.* **2020**, *594*, 117339. [CrossRef]
25. Wang, M.; Wang, J.; Jiang, J. Membrane Fouling: Microscopic Insights into the Effects of Surface Chemistry and Roughness. *Adv. Theory Simul.* **2022**, *5*, 2100395. [CrossRef]
26. Alsawaftah, N.; Abuwatfa, W.; Darwish, N.; Husseini, G. A Comprehensive Review on Membrane Fouling: Mathematical Modelling, Prediction, Diagnosis, and Mitigation. *Water* **2021**, *13*, 1327. [CrossRef]
27. Lim, M.Z.Y.; Chong, W.C.; Lau, W.J.; Koo, C.H. Performance of Thin Film Composite Membranes for Ammonium Removal and Reuse of Ammonium-Enriched Solution for Plant Growth. *Water Sci. Technol. Water Supply* **2021**, *21*, 318–330. [CrossRef]
28. Yang, X.; Yan, L.; Wu, Y.; Liu, Y.; Shao, L. Biomimetic Hydrophilization Engineering on Membrane Surface for Highly-Efficient Water Purification. *J. Membr. Sci.* **2019**, *589*, 117223. [CrossRef]

29. Gao, L.; Wang, H.; Zhang, Y.; Wang, M. Nanofiltration Membrane Characterization and Application: Extracting Lithium in Lepidolite Leaching Solution. *Membranes* **2020**, *10*, 178. [CrossRef]
30. He, M.; Yuan, T.; Dong, W.; Li, P.; Jason Niu, Q.; Meng, J. High-Performance Acid-Stable Polysulfonamide Thin-Film Composite Membrane Prepared via Spinning-Assist Multilayer Interfacial Polymerization. *J. Mater. Sci.* **2019**, *54*, 886–900. [CrossRef]
31. Kaya, C.; Jarma, Y.A.; Muhidin, A.M.; Güler, E.; Kabay, N.; Arda, M.; Yüksel, M. Seawater Desalination by Using Nanofiltration (NF) and Brackish Water Reverse Osmosis (BWRO) Membranes in Sequential Mode of Operation. *J. Membr. Sci. Res.* **2020**, *6*, 40–46. [CrossRef]
32. Idarraga-Mora, J.A.; Lemelin, M.A.; Weinman, S.T.; Husson, S.M. Effect of Short-Term Contact with C1–C4 Monohydric Alcohols on the Water Permeance of MPD-TMC Thin-Film Composite Reverse Osmosis Membranes. *Membranes* **2019**, *9*, 92. [CrossRef] [PubMed]
33. Sienkiewicz, A.; Krasucka, P.; Charmas, B.; Stefaniak, W.; Goworek, J. Swelling Effects in Cross-Linked Polymers by Thermogravimetry. *J. Therm. Anal. Calorim.* **2017**, *130*, 85–93. [CrossRef]
34. Wang, Z. Cross-Linked Aromatic Polyamide Membranes. In *Encyclopedia of Membranes*; Springer: Berlin/Heidelberg, Germany, 2015; pp. 1–2.
35. RisingSun Membrane NF Flat Sheet Membranes. Available online: https://www.risingsunmembranes.com/flat-sheet-membranes/nf-flat-sheet-membranes/nf1-nf2-nf3-flat-sheet-membrane-mwco-150-400.html (accessed on 27 February 2023).
36. Suez (GE) Flat Sheet Membrane, Duracid NF. Available online: https://www.sterlitech.com/nanofiltration-nf-membrane-ymkhsp475.html?wtx=GAW&gclid=Cj0KCQjwuLShBhC_ARIsAFod4fK6ST5rdTXYUwTdzDB8tsnPTiAJudZAZIHctLafbJDPUkLOIL0d4QsaArodEALw_wcB (accessed on 27 February 2023).
37. Bargeman, G. Recent Developments in the Preparation of Improved Nanofiltration Membranes for Extreme PH Conditions. *Sep. Purif. Technol.* **2021**, *279*, 119725. [CrossRef]
38. Gao, M.; Gong, X.; Lv, M.; Song, W.; Ma, X.; Qi, Y.; Wang, L. Effect of Temperature and PH on the Sorption of Dibutyl Phthalate on Humic Acid. *Water Air Soil Pollut.* **2016**, *227*, 55. [CrossRef]
39. Dada, E.O.; Ikeh, R.K. Phthalate and Metal Concentrations in Drinking Water in Lagos, Nigeria. *J. Health Pollut.* **2018**, *8*, 180603. [CrossRef]

membranes

MDPI

Article

Anion Effect on Forward Osmosis Performance of Tetrabutylphosphonium-Based Draw Solute Having a Lower Critical Solution Temperature

Jihyeon Moon and Hyo Kang *

BK-21 Four Graduate Program, Department of Chemical Engineering, Dong-A University, 37, Nakdong-Daero 550 Beon-gil, Saha-gu, Busan 49315, Republic of Korea
* Correspondence: hkang@dau.ac.kr; Tel.: +82-51-200-7720; Fax: +82-51-200-7728

Abstract: The applicability of ionic liquids (ILs) as the draw solute in a forward osmosis (FO) system was investigated through a study on the effect of the structural change of the anion on the FO performance. This study evaluated ILs composed of tetrabutylphosphonium cation ($[P_{4444}]^+$) and benzenesulfonate anion ($[BS]^-$), para-position alkyl-substituted benzenesulfonate anions (*p*-methylbenzenesulfonate ($[MBS]^-$) and *p*-ethylbenzenesulfonate ($[EBS^-]$), and methanesulfonate anion ($[MS]^-$). The analysis of the thermo-responsive properties suggested that the $[P_{4444}]$[MBS] and $[P_{4444}]$[EBS] ILs have lower critical solution temperatures (LCSTs), which play a beneficial role in terms of the reusability of the draw solute from the diluted draw solutions after the water permeation process. At 20 wt% of an aqueous solution, the LCSTs of $[P_{4444}]$[MBS] and $[P_{4444}]$[EBS] were approximately 36 °C and 25 °C, respectively. The water flux and reverse solute flux of the $[P_{4444}]$[MBS] aqueous solution with higher osmolality than $[P_{4444}]$[EBS] were 7.36 LMH and 5.89 gMH in the active-layer facing the draw solution (AL-DS) mode at osmotic pressure of 25 atm (20 wt% solution), respectively. These results indicate that the $[P_{4444}]^+$-based ionic structured materials with LCST are practically advantageous for application as draw solutes.

Keywords: ionic liquid; forward osmosis; draw solution; water treatment; lower critical solution temperature

1. Introduction

Since water shortages have become a critical global problem owing to the increasing demand for clean water, a supply of adequate water is crucial to sustain public health and national prosperity [1–3]; therefore, the development of water treatment technologies is essential. Among the various water purification technologies that have been investigated, forward osmosis (FO) has been considered a feasible energy-efficient process [4–6]. Unlike conventional pressure-driven membrane processes, the FO membrane process is activated by the chemical potential difference between a feed solution and draw solution, which are placed on different sides of a semipermeable membrane. This natural phenomenon allows the transport of water molecules across the membrane, leading to the dilution of the draw solution in the FO process [7,8]. FO processes that do not require external pressure exhibit low fouling tendencies and high fouling reversibility of membranes, thereby increasing membrane life [9–11]. However, the industry has not fully accommodated the FO process because an additional recovery process to separate water from the diluted draw solution is required, which is energy-intensive [12]. The design or selection of appropriate draw solutions with high water-drawing ability and easy recovery is key in improving the FO process [13,14].

Researchers have proposed two types of draw solutes such as non-responsive [15–17] and responsive [18–20]. To achieve easy recovery, various responsive draw solutes have been proposed. The responsive draw solutes undergo a change in water solubility upon exposure to external environmental parameters, such as temperature [18,21], pH [22], CO_2 [23,24], and electric and magnetic fields [25–28]. Thermo-responsive draw solutes have especially attracted attention for recovering draw solutes and clean water because of their simplicity and low-energy usage in the recovery process via a simple temperature control using waste or geothermal heat [29–31]. Thermo-responsive draw solutes can exhibit one of the following two types of phase transition: upper critical solution temperature (UCST) and lower critical solution temperature (LCST). With the UCST-type, the compatibility of the draw solute and water increases upon heating, while the water solubility of the LCST-type draw solute decreases, implying easy recovery.

Various thermo-responsive draw solutes have been explored, which include hydrogels [32,33], homopolymers [18,34], oligomers [20], and copolymers [21]. Recently, thermo-responsive ILs have been considered promising draw solutes because ILs can dissociate in water, leading to osmolality [35,36]. ILs have several distinct characteristics, such as high ionic conductivity, thermal stability, tunable polarity, and negligible volatility [37–40]. Moreover, the physicochemical properties can be tailored by altering the ion species [41,42]. ILs are defined as compounds composed of a positively charged cation and a negatively charged anion. In the case of cation–anion combinations of ILs that show thermo-responsive phase behavior, typically, the anions include both organic and inorganic species, but cations include usually organic species such as imidazolium, pyridinium, phosphonium, and ammonium [43,44]. Several thermo-responsive ILs were introduced as the draw solute at 20–60 °C, such as tetrabutylammonium 2,4,6-trimethylbenzenesulfonate ([N_{4444}]2,4,6-$MeBnSO_3$) [45], tetrabutylphosphonium trifluoroacetate ([P_{4444}]CF_3COO) [46], 1-butyl-3-methylimidazolium tetrafluoroborate ([Bmim][BF_4]) [47], tetraethylammonium bromide ([N_{2222}]Br) [48], betaine bis(trifluoromethylsulfonyl)imide ([Hbet][Tf_2N]) [49], and poly(4-vinylbenzyltributylammonium hexanesulfonate) (P[VBTBA][HS]) [50]. These examples refer to the situation where the thermo-responsive ILs are available for draw solute in the FO system. The relationship between the structure and phase behavior of LCST-type ILs in an aqueous solution has also been investigated. Many reports show that the main factor affecting the LCST behavior of the ILs in water is the total hydrophobicity of the ILs, which can be tuned by changing the hydrophobic moieties of the cation or anion [51–56]. Therefore, a structure–property correlation study of ILs can support the design of future draw solutes with water-drawing and simple separation abilities. For example, the effect of the hydrophobic alkyl chain groups on a specific property has been studied using several series of ILs such as quaternary phosphonium-based ILs paired with different anions, including tetrachloroferrate (III), amino acid, and sulfonate types [55–57]. Among various ILs, phosphonium-based ILs have been reported to have more thermal and chemical stability [58–60]. The phosphonium-based ILs become attractive materials due to their competitive cost of synthesis and higher productivity at the industrial production level as well as their stability [61,62]. According to previous literature, the ILs based on phosphonium with shorter alkyl chain lengths have shown low toxicity toward several human microorganisms and they present the potential for bioprocessing applications [63]. Therefore, we investigated the usability of tetrabutylphosphonium-based ILs in combination with benzenesulfonate which has a sulfonate group attached to the phenyl ring, a simple hydrophobic group. The hydrophobicity of benzenesulfonate anions can be tuned by attaching the extra methyl group to the phenyl ring. In addition, benzenesulfonate also has the advantage of low production cost and high thermal stability [64,65].

This study investigated the water-drawing and separation abilities of a series of tetrabutylphosphonium ([P_{4444}]$^+$)-based ILs formed by benzenesulfonate anion ([BS]$^-$), para-position alkyl-substituted benzenesulfonate anions (*p*-methylbenzenesulfonate ([MBS]$^-$) and *p*-ethylbenzenesulfonate ([EBS$^-$])), and methanesulfonate anion ([MS]$^-$) with different hydrophobicity. The effect of changing the hydrophobic moiety of anions in ILs on their osmolality and LCST behavior was systematically observed. These investigations provide a guide for the development of draw solutes in improving the efficiency of the FO system.

2. Materials and Methods

2.1. Materials

Tetrabutylphosphonium bromide, $[P_{4444}]Br$ (99.0%), sodium benzenesulfonate, Na[BS] (>96.0%), sodium *p*-methylbenzenesulfonate (>90.0%) (Na[MBS]), sodium *p*-ethylbenzenesulfonate, Na[EBS] (>98.0%), and tetrabutylphosphonium methanesulfonate, $[P_{4444}][MS]$ (>98.0%) were purchased from Tokyo Chemical Industry Co., Ltd. (Tokyo, Japan). Dichloromethane was purchased from Daejung Chemicals and Metals Co., Ltd. (Sinan, Republic of Korea).

2.2. Preparation of Phosphonium-Based ILs

Tetrabutylphosphonium benzenesulfonate ($[P_{4444}][BS]$), tetrabutylphosphonium *p*-methylbenzenesulfonate ($[P_{4444}][MBS]$), and tetrabutylphosphonium *p*-ethylbenzenesulfonate ($[P_{4444}][EBS]$) were prepared by mixing $[P_{4444}]Br$ (3.39 g, 10 mmol) and sodium benzene derivatives, Na[BS] (3.60 g, 20 mmol), Na[MBS] (3.88 g, 20 mmol), and Na[EBS] (4.16 g, 20 mmol) in 20 mL of distilled water at 25 °C for 24 h. The products, $[P_{4444}][BS]$, $[P_{4444}][MBS]$, and $[P_{4444}][EBS]$ were extracted with dichloromethane and purified with distilled water. The resulting products were obtained after the removal of dichloromethane by evaporation and drying in a vacuum oven at 40 °C for 48 h. In contrast, tetrabutylphosphonium methanesulfonate ($[P_{4444}][MS]$) was used as received without any further processing. To identify the structure of the ILs, proton nuclear magnetic resonance (^{1}H-NMR) spectroscopy was used.

^{1}H-NMR of $[P_{4444}][BS]$ (400 MHz, D_2O, δ/ppm): δ = 0.84–1.01 (t, 12H, -P^+-CH_2-CH_2-CH_2-CH_3), 1.35–1.66 (m, 16H, -P^+-CH_2-CH_2-CH_2-CH_3), 2.03–2.28 (t, 8H, -P^+-CH_2-CH_2-CH_2-CH_3, 7.46–7.69, 7.72–7.92 (s, 5H, *PhH*-SO_3^-).

^{1}H-NMR of $[P_{4444}][MBS]$ (400 MHz, D_2O, δ/ppm): δ = 0.83–1.03 (t, 12H, -P^+-CH_2-CH_2-CH_2-CH_3), 1.35–1.67 (m, 16H, -P^+-CH_2-CH_2-CH_2-CH_3), 2.06–2.25 (m, 3H, -CH_3-Ph-SO_3^-), 2.31–2.49 (t, 8H, -P^+-CH_2-CH_2-CH_2-CH_3), 7.29–7.45, 7.61–7.77 (s, 4H, CH_3-*PhH*-SO_3^-).

^{1}H-NMR of $[P_{4444}][EBS]$ (400 MHz, D_2O, δ/ppm): δ = 0.72–1.07 (t, 12H, -P^+-CH_2-CH_2-CH_2-CH_3), 1.12–1.36 (t, 3H, -CH_3-CH_2-Ph-), 1.36–1.83 (m, 16H, -P^+-CH_2-CH_2-CH_2-CH_3), 1.90–2.51 (t, 8H, -P^+-CH_2-CH_2-CH_2-CH_3), 2.61–2.88 (m, 2H, -CH_3-CH_2-Ph-), 7.30–7.52 (d, 4H, CH_3-CH_2-*PhH*-SO_3^-), 7.62–7.81 (d, 4H, CH_3-CH_2-*PhH*-SO_3^-).

^{1}H-NMR of $[P_{4444}][MS]$ (400 MHz, D_2O, δ/ppm): δ = 0.79–1.07 (t, 12H, -P^+-CH_2-CH_2-CH_2-CH_3), 1.29–1.75 (m, 16H, -P^+-CH_2-CH_2-CH_2-CH_3), 1.99–2.31 (t, 8H, -P^+-CH_2-CH_2-CH_2-CH_3), 2.70–2.90 (s, 3H, CH_3-SO_3-).

2.3. FO Performance

The water flux, which represents the FO performance, was measured using a lab-scale FO apparatus connected to custom-made, L-shaped glass tubes. The membrane (Hydration Technologies Inc. (HTI, Albany, OR, USA), commercial thin film composite membrane) was inserted between two glass tubes composed of a 3.3 cm diameter fastening aluminum clamp and FO was carried out to evaluate its performance. The parameter of the FO performance or water flux (J_w) was measured in the active-layer facing the draw solution (AL-DS) mode and the active-layer facing the feed solution (AL-FS) mode, respectively. Distilled water was added as a feed solution to one of the tubes and $[P_{4444}][MBS]$ was added as a draw solution to the other. The measurement was performed by stirring with a solenoid (OCTOPUS CS-4, AS ONE, Osaka, Japan) at 25 ± 1 °C for 20 min.

We calculated the water flux, J_w, by converting the height difference to the volumetric change of the draw solution using the following Equation (1):

$$J_w = \frac{\Delta V}{A \Delta t} \tag{1}$$

where Δt $[h]$ is the handling time of the experiment of the FO process, ΔV $[L]$ is the variation in volume of the draw solution, and A $[m^2]$ is the effective membrane area, which is 4.15×10^{-4} m^2.

The reverse solute flux (J_s, g m^{-2} h^{-1}, gMH) represents the quantity of the permeated draw solute across the FO membrane to the feed solution. We calculated the reverse solute flux by comparing the conductivity difference of the feed solution before and after FO using the following Equation (2):

$$J_s = \frac{\Delta(CV)}{A\Delta t} \quad (2)$$

where ΔC $[mg/L]$ is the concentration change and ΔV $[L]$ is the variation in volume of the feed solution before and after FO.

2.4. Instruments

To confirm the molecular structure of the prepared ILs, ^{1}H-NMR spectroscopy (MR400 DD2 NMR, Agilent Technologies, Inc., Santa Clara, CA, USA) and Fourier transform infrared (FT-IR) spectroscopy (Nicolet iS20, Thermo Fisher Scientific Inc., Waltham, MA, USA) were used. A conductivity meter (Seven2CO pro, METTLER TOLEDO Inc., Columbus, OH, USA) was used to measure the conductivity. An osmometer (SEMI-MICRO OSMOMETER K-7400, KNAUER Wissenschaftliche Geräte GmbH Co., Berlin, Germany) was used to obtain the osmotic pressure of the IL aqueous solutions. An ultraviolet-visible (UV-Vis) spectrophotometer (EMC-11D-V, EMCLAB Instruments GmbH Co., Duisburg, Germany) fitted with a temperature controller (TC200P, Misung Scientific Co., Ltd., Yangju, Republic of Korea) was used to confirm the phase separation temperature of the ILs in water. The measurement of the contact angle was performed using a Krüss DSA10 (KRÜSS Scientific Instruments Inc., Hamburg, Germany) contact angle analyzer equipped with drop shape analysis software after deposing the water and aqueous solution of IL droplets on the FO membrane surface. The average volume of the droplets was 5 μL. The contact angles of each solution were determined four or more times.

3. Results and Discussion

3.1. Synthesis and Characterization of Phosphonium-Based ILs

The tetrabutylphosphonium-based ILs with benzenesulfonate derivative anions were prepared via anion exchange of tetrabutylphosphonium bromide ([P_{4444}]Br) from bromide to [BS], [MBS], and [EBS], as illustrated in Figure 1. The structures of [P_{4444}][BS], [P_{4444}][MBS], and [P_{4444}][EBS] were determined using ^{1}H-NMR and attenuated total reflectance FT-IR (ATR FT-IR) analysis. The ^{1}H-NMR spectra of the [P_{4444}][BS], [P_{4444}][MBS], [P_{4444}][EBS], and [P_{4444}][MS] are depicted in Figure 2a–d. The successful preparation of [P_{4444}][BS], [P_{4444}][MBS], [P_{4444}][EBS], and [P_{4444}][MS] was confirmed by calculating the integral ratio for each region of the protons of the alkyl groups. Furthermore, the anion exchange of [P_{4444}][BS], [P_{4444}][MBS], and [P_{4444}][EBS] was confirmed by comparison of the respective peaks of the resulting products with those of Na[BS], Na[MBS], and Na[EBS] using FT-IR spectroscopy, as illustrated in Figure 3a–c. [P_{4444}][MS] was also confirmed (Figure 3d). All resulting products exhibit two characteristic peaks corresponding to asymmetrical and symmetrical stretching vibrations of C–H in the alkyl groups at 2962–2957 cm^{-1} and 2870–2867 cm^{-1}, respectively. Although the absorption at 1600–1500 cm^{-1} should be assigned to the stretching of the C=C bands in the benzene ring, the coarse spectra make it difficult to trace the variation in this region [66]. However, bands at 1200–1193 cm^{-1} and 1120–1115 cm^{-1} correspond to the asymmetrical and symmetrical stretching of S=O, which makes it possible to follow the anion variations. Therefore, the absorption of C–H and S=O bands brings particular convenience to the analysis of cations and anions in this system.

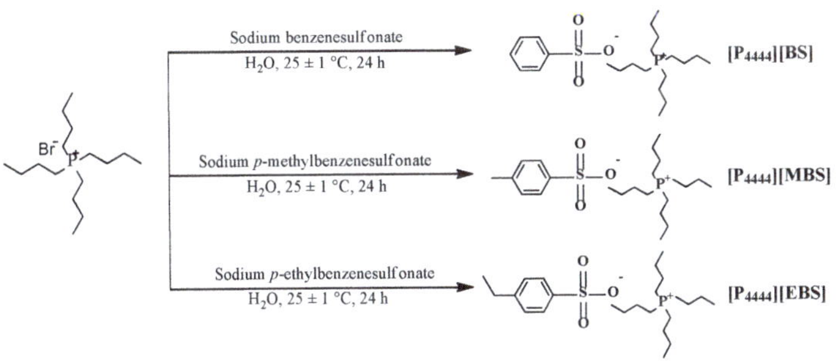

Figure 1. Preparative scheme for $[P_{4444}][BS]$, $[P_{4444}][MBS]$, and $[P_{4444}][EBS]$.

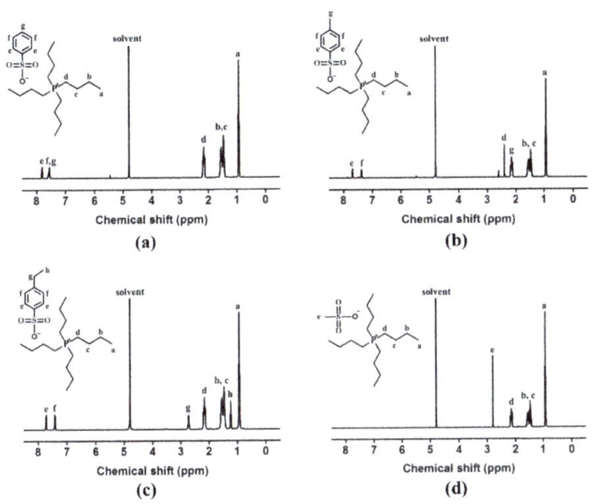

Figure 2. Proton nuclear magnetic resonance (^{1}H-NMR) spectra of (**a**) $[P_{4444}][BS]$, (**b**) $[P_{4444}][MBS]$, (**c**) $[P_{4444}][EBS]$, and (**d**) $[P_{4444}][MS]$.

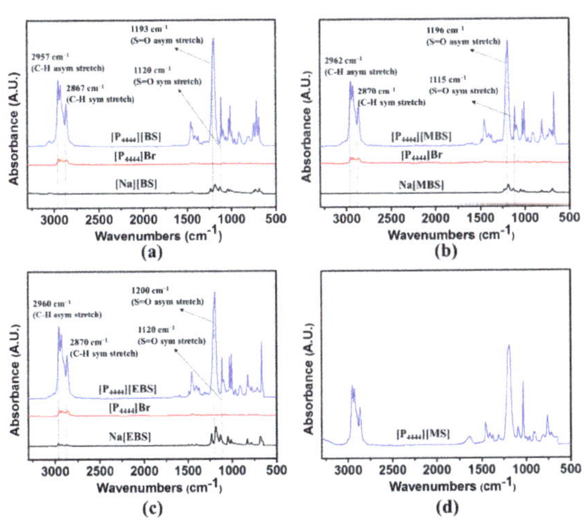

Figure 3. Fourier-transform infrared (FT-IR) spectra of (**a**) $[P_{4444}][BS]$, (**b**) $[P_{4444}][MBS]$, (**c**) $[P_{4444}][EBS]$, and (**d**) $[P_{4444}][MS]$.

3.2. Electrical Conductivity

As an indicator of the ion dissociation ability, electrical conductivity is affected by the ionic species, which influence the charge carrier and ion mobility. Typically, the large ions cause ion aggregation/pairing and have more interfacial area between the ion and solvent, increasing the hydrodynamic resistance, and resulting in limited ion mobility [67,68]. In addition, the conductivity is related to the osmolality of ILs, which is largely dependent on ion dissociation [69]. Thus, the conductivities of the [P_{4444}][MS], [P_{4444}][BS], [P_{4444}][MBS], and [P_{4444}][EBS] aqueous solutions were measured at the concentration range of 5–20 wt%. The conductivities of the [P_{4444}][MS], [P_{4444}][BS], [P_{4444}][MBS], and [P_{4444}][EBS] aqueous solutions were approximately 9187, 5460, 4510, and 3710 μS/cm, respectively, at a concentration of 10 wt%. The conductivities of [P_{4444}][MS], [P_{4444}][BS], [P_{4444}][MBS], and [P_{4444}][EBS] aqueous solutions increased to approximately 12,900, 7037, 5436, and 4009 μS/cm, respectively, when their concentration was increased to 20 wt%. As shown in Figure 4, the conductivities of the IL aqueous solutions increase with increasing concentration owing to increasing numbers of mobile ions in the aqueous solution in the range of 5–20 wt% [70–74]. Furthermore, the conductivity decreases in the order [P_{4444}][MS] < [P_{4444}][BS] < [P_{4444}][MBS] < [P_{4444}][EBS], which is also the order of increasing ion size at any concentration of ILs. As aforementioned, the charge carriers and ionic mobility of the ILs containing large ion sizes decrease, resulting in a decrease in ionic conductivity. In addition, aromatic substituents enhance the van der Waals interactions, which lead to low ion conductivities [75]. Therefore, tetrabutylphosphonium-based ILs containing small anions show a relatively high conductivity.

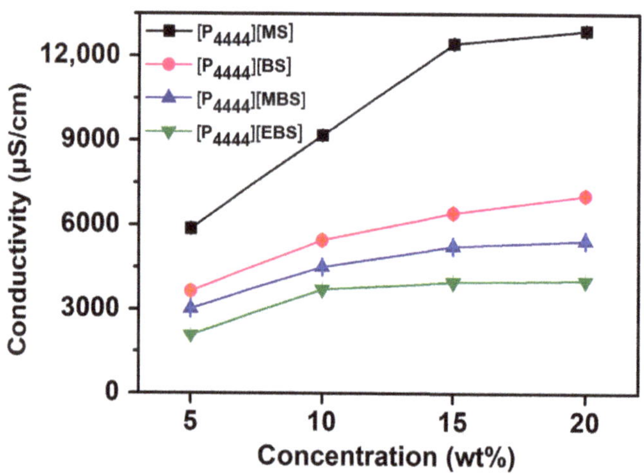

Figure 4. Electrical conductivities of [P_{4444}][MS], [P_{4444}][BS], [P_{4444}][MBS], and [P_{4444}][EBS] aqueous solutions according to the concentration of solution.

3.3. Osmotic Pressure

The transport of the water molecules across the membrane is activated by the high osmotic pressure of the draw solution compared to that of the feed solution; thus, it can be used as the driving force of the FO system [30,76]. As a colligative property, the increase in the concentration of the draw solution enhances its osmotic pressure [77]. To investigate the applicability as a draw solute, the osmolality values of the [P_{4444}][MS], [P_{4444}][BS], [P_{4444}][MBS], and [P_{4444}][EBS] aqueous solutions were measured at increasing IL concentrations from 5 to 20 wt% using the freezing point depression method. The osmolality values could be converted to the osmotic pressure via the van't Hoff equation using the temperature (T = 297 K) and density of the solution (density = 1 g/mL), as depicted in Figure 5. The osmotic pressures of the [P_{4444}][MS], [P_{4444}][BS], [P_{4444}][MBS], and $[P_{4444}][EBS]$ aqueous solutions increased from 17, 14, 11, and 9 atm to 34, 33, 25, and 18 atm, respectively, when their concentration increased from 10 to 20 wt%. As a colligative property, the osmotic pressure is correlated with the degree of ion dissociation in water, which is largely dependent on ion solvation [78,79]. In addition to cation–anion interactions, solvent–ion, and solvent–solvent interactions determine how well the ions are solvated, and consequently the degree of ion dissociation [80]. Hydrogen bonding, particularly between solvent and anion, plays a key role in ion solvation [81]. The extent of ion dissociation decreases with increasing hydrophobicity of the anion owing to the increased size of the hydrophobic moiety such as a benzene derivative; this requires a greater number of water molecules for solvation based on anion size [82–84]. Thus, ILs with highly hydrophilic $[P_{4444}][MS]$, containing $[MS]^-$ anions, which is considered a factor in their high degree of ion dissociation, show a higher osmotic pressure at all concentrations. Moreover, the addition of the hydrophobic methyl group to the benzenesulfonate anion decreases the solvation of the benzenesulfonate-based ILs with water; thus, the osmotic pressure decreases in the order: $[P_{4444}][BS] > [P_{4444}][MBS] > [P_{4444}][EBS]$. Therefore, at any concentration of the ILs, the osmotic pressure increases in the order of $[P_{4444}][MS] > [P_{4444}][BS] > [P_{4444}][MBS] > [P_{4444}][EBS]$, which is also the order of increasing hydrophilicity. The $[P_{4444}][MS]$, $[P_{4444}][BS]$, and $[P_{4444}][MBS]$ aqueous solutions generate osmotic pressures of 34–25 atm at a concentration of 20 wt%, as compared to 26 atm generated by traditional inorganic salts especially NaCl with 3.5 wt%. Although the osmotic pressures of the $[P_{4444}]$-based ILs aqueous solutions are lower than NaCl, they exhibit potential as draw solution for brackish water treatment and food processing [85]. In addition, overall, the $[P_{4444}]$-based ILs have not only good solubility and a high degree of dissociation at a concentration of 20 wt% but also thermal recovery properties.

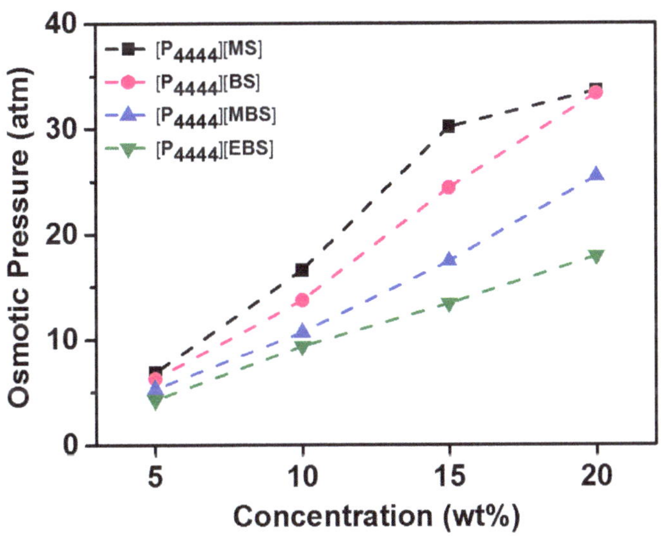

Figure 5. Osmotic pressures of $[P_{4444}][MS]$, $[P_{4444}][BS]$, $[P_{4444}][MBS]$, and $[P_{4444}][EBS]$ aqueous solutions according to the concentration of solution.

3.4. Thermo-Responsive Property

Thermo-responsive ILs facilitate the efficient separation from the diluted draw solution into clean water and draw solute [86]. The LCST is the phase separation temperature at which the ILs with LCST become a heterogeneous state in an aqueous solution above the LCST. The LCST behavior is attributed to the following equation, where ΔG_{mix} is the mixing free energy, ΔH_{mix} is the mixing enthalpy, and ΔS_{mix} is the mixing entropy [87–89].

$$\Delta G_{mix} = \Delta H_{mix} - T\Delta S_{mix} \quad (3)$$

In the aqueous solution of an IL system, the interaction between ions and water, such as hydrogen bonding plays an important role in the LCST-type phase behavior of the IL. The mixing enthalpy is negative at low temperature because the hydrogen bonding makes a negative ΔS_{mix} and a miscible phase between the component species, thus ΔG_{mix} must be negative. Upon heating to a temperature above LCST, ΔG_{mix} changes from negative to positive leading to phase separation due to the breaking of the hydrogen bonding. In our system, the $[P_{4444}]$[MBS] and $[P_{4444}]$[EBS] aqueous solutions exhibit an LCST-type phase transition, as can be seen in Figure 6. Below its LCST, the hydrogen bonding interaction between water and the $[P_{4444}]$[MBS] and $[P_{4444}]$[EBS] is dominant. But above its LCST, the hydrogen bonding interaction is weakened and the hydrophobic interaction between $[P_{4444}]^+$ and benzenesulfonate derivatives anions ($[MBS]^-$ and $[EBS]^-$) becomes dominant, thereby inducing the phase separation of $[P_{4444}]$[MBS] and $[P_{4444}]$[EBS]. In contrast, there were no noticeable changes in transmittance for the $[P_{4444}]$[MS] and $[P_{4444}]$[BS] aqueous solutions within the entire analyzed temperature range. Because the main factor affecting the LCST behavior of the ILs in water is the total hydrophobicity of the ILs [56]. When the IL composed of cation and anion is highly hydrophilic, IL is miscible in water and does not exhibit phase separation. The LCST of the $[P_{4444}]$[MBS] and $[P_{4444}]$[EBS] aqueous solutions were determined by UV-Vis spectroscopy and temperature controller at 550 nm upon heating to 90 °C. The transmittance versus temperature curves of the $[P_{4444}]$[MBS] and $[P_{4444}]$[EBS] aqueous solutions mark changes at each LCST for concentrations of 10, 15, and 20 wt%, as shown in Figure 6. As the concentration of the $[P_{4444}]$[MBS] and $[P_{4444}]$[EBS] aqueous solution increases, the LCST decreases. The LCSTs of the $[P_{4444}]$[MBS] aqueous solutions are approximately 54, 39, and 36 °C, at each concentration, respectively. Additionally, the respective LCSTs of $[P_{4444}]$[EBS] are approximately 31, 28, and 19 °C. As a result of the thermo-responsive phase separation behavior of the prepared ILs in an aqueous solution, the $[P_{4444}]$[MBS] and $[P_{4444}]$[EBS] aqueous solutions could be easily separated from the draw solute and pure water by heating and cooling, requiring minimal recovery energy during FO owing to their thermo-responsiveness.

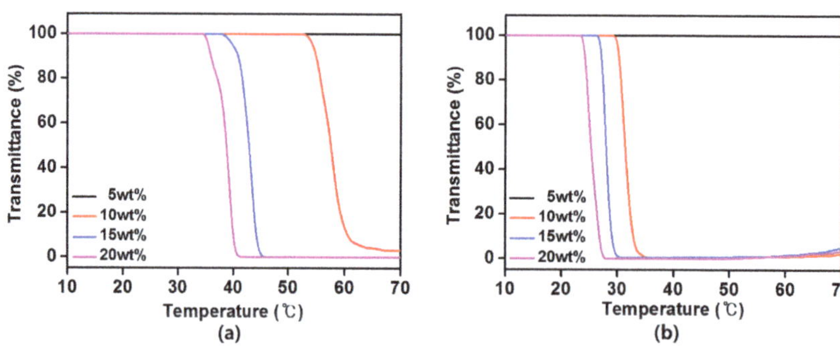

Figure 6. Transmittance curve of (**a**) $[P_{4444}]$[MBS] and (**b**) $[P_{4444}]$[EBS] aqueous solutions according to the temperature.

3.5. Contact Angle of $[P_{4444}]$[MBS] on the FO Membrane

The wettability of the membrane surface is a macroscopic representation of the interfacial interaction between fluid and membrane surface and plays an important role in the membrane performance [90–93]. The surface wetting of the membrane is usually characterized by the contact angle. The contact angles of the $[P_{4444}]$[MBS] aqueous solutions with various concentrations and distilled water on the surface of the active layer in the FO membrane were measured to understand the effect of the draw solution ($[P_{4444}]$[MBS] aqueous solution) on the wettability of the membrane. As shown in Figure 7, the average contact angle of distilled water is lower than that of $[P_{4444}]$[MBS] aqueous solution at all concentrations on the membrane surface and follows the following trend: distilled water > 5 wt% $[P_{4444}][MBS]$ > 10 wt% $[P_{4444}][MBS]$ > 15 wt% $[P_{4444}][MBS]$ > 20 wt% $[P_{4444}][MBS]$. The distilled water contact angle was 75°. The contact angles of the $[P_{4444}][MBS]$ aqueous solutions were 54, 49, 39, and 21°, respectively, at concentrations of 5, 10, 15, and 20 wt%, respectively. The contact angle of $[P_{4444}][MBS]$ aqueous solution on the membrane becomes smaller with increasing concentration. This means that the wettability of the FO membrane was enhanced on decreasing the contact angle of the $[P_{4444}][MBS]$ aqueous solution from 54° to 21°.

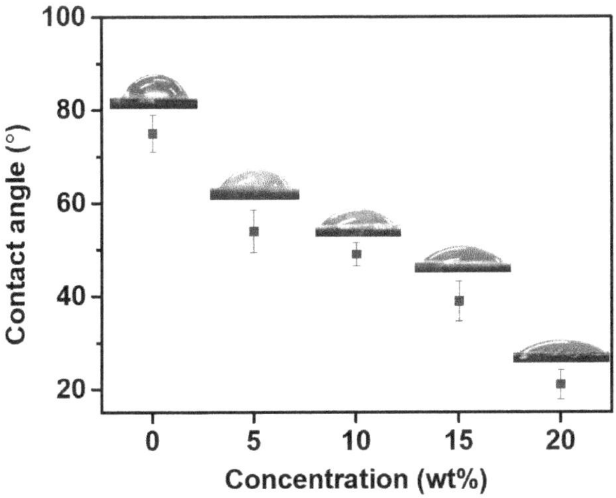

Figure 7. Contact angles of $[P_{4444}][MBS]$ aqueous solution at concentrations of 0 (distilled water), 5, 10, 15, and 20 wt% on the FO membrane.

3.6. Water Flux (J_w) and Reverse Solute Flux (J_s) of $[P_{4444}][MBS]$

The water flux and reverse solute flux of $[P_{4444}][MBS]$, which has a high osmolality potential among the thermo-responsive ILs, were measured to evaluate $[P_{4444}][MBS]$ as a draw solute at different orientations of the membrane placed between the two connected glass tubes; the glass tubes were filled with distilled water and $[P_{4444}][MBS]$ aqueous solution, respectively. At 25 ± 1 °C, the water fluxes and reverse solute fluxes of the $[P_{4444}][MBS]$ aqueous solutions were measured in the AL-DS and AL-FS modes when the concentration of ILs varied from 5 to 20 wt%. The concentrations of 5, 10, 15, and 20 wt% are represented as 5, 11, 17, and 25 atm, respectively. The water fluxes of the $[P_{4444}][MBS]$ aqueous solutions were 1.67, 3.27, 4.64, and 7.36 LMH in the AL-DS mode at osmotic pressures of 5, 11, 17, and 25 atm, respectively, and 1.65, 2.80, 3.45, and 4.85 LMH in the AL-FS mode, respectively. The reverse solute flux is the quantity of the permeated draw solute across the FO membrane to the feed water and was calculated by comparing the amount of total dissolved solids (TDS) of the distilled water (feed solution) before and after FO [94]. The reverse solute fluxes of the $[P_{4444}][MBS]$ aqueous solutions were 1.36, 1.47, 3.77, and 5.89 gMH in the AL-DS mode at 5, 11, 17, and 25 atm, respectively, and 1.14, 1.25, 2.45, and 2.67 gMH in the AL-FS mode, respectively, for the same condition. We observed that the water flux increases when the osmotic pressure of IL increases. This trend indicates that the osmotic pressure induces water permeation for the FO process. Figure 8 shows that the water fluxes are higher for the AL-DS mode than for the AL-FS mode at all $[P_{4444}][MBS]$ aqueous solutions, which is attributed to a significant dilutive internal concentration polarization (ICP) effect in the AL-FS mode [95]. This causes the dilution of the draw solution, which reduces the effective osmotic pressure across the membrane. When the feed solution is distilled water, the ICP effect is negligible in the AL-DS mode [96,97].

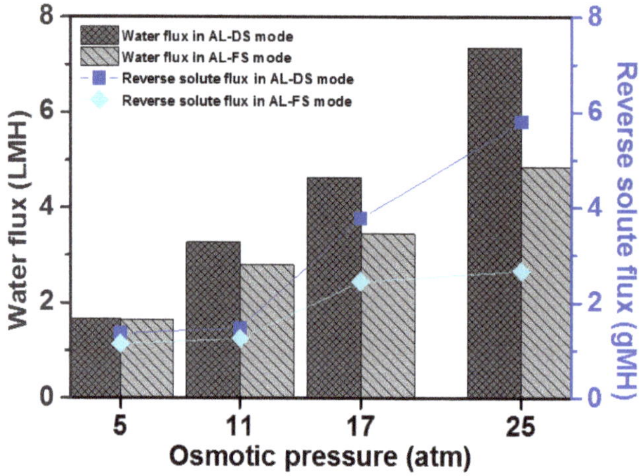

Figure 8. Water fluxes and reverse solute fluxes of $[P_{4444}][MBS]$ aqueous solutions according to the osmotic pressure in AL-DS mode and AL-FS mode at a temperature of 25 ± 1 °C during FO process.

3.7. Recyclability Study of $[P_{4444}][MBS]$

To explore the recyclable property of the $[P_{4444}][MBS]$, the FO performance of 20 wt% aqueous solutions of the $[P_{4444}][MBS]$ was repeated four times when distilled water was used as the feed solution. The thermal treatment method was used to obtain the recycled $[P_{4444}][MBS]$. As shown in Figure 9a,b, the osmotic pressure and LCST of the $[P_{4444}][MBS]$ were measured to confirm the recyclability of $[P_{4444}][MBS]$ at the fourth run. The osmotic pressure values at the 4th run were almost the same as that of pristine $[P_{4444}][MBS]$, while the LCST value slightly increased after the 4th run. These recycling results clearly show that $[P_{4444}][MBS]$ can be easily recycled with relatively low energy consumption without significant loss.

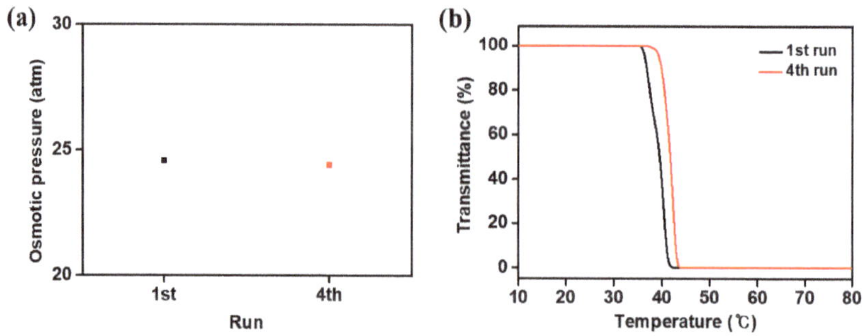

Figure 9. (**a**) Osmotic pressures and (**b**) LCST results using 20 wt% $[P_{4444}][MBS]$ aqueous solution as a draw solution and distilled water as a feed solution. From the 1st to the 4th run, the recovered $[P_{4444}][MBS]$ from the previous run was used.

4. Conclusions

The water-drawing ability and LCST-type phase transition behavior of IL draw solutes composed of tetrabutylphosphonium cation ($[P_{4444}]^+$) and anions with different hydrophobicity ($[BS]^-$, $[MBS]^-$, $[EBS]^-$, and $[MS]^-$) were described in this article. The total hydrophobicity of the ILs affects the osmolality and LCST-type phase behavior of the draw solution. The hydrophilicity of the draw solutions followed the order $[P_{4444}][MS]$ > $[P_{4444}][BS]$ > $[P_{4444}][MBS]$ > $[P_{4444}][EMS]$ at all concentrations. ILs with high hydrophilicity have a high osmolality potential and LCST. During the phase transition behavior, LCSTs were observed from the [P_{4444}][MBS] and [P_{4444}][EBS] aqueous solutions. The LCSTs of the 20 wt% [P_{4444}][MBS] and [P_{4444}][EBS] aqueous solutions were approximately 36 °C and 25 °C, respectively. In the FO performance test, at an osmotic pressure of 25 atm, the water fluxes and reverse solute fluxes of the thermo-responsive [P_{4444}][MBS] aqueous solution that exhibited higher osmolality compared to [P_{4444}][EBS] were 7.36 LMH and 5.89 gMH in the AL-DS, and 4.85 LMH and 2.67 gMH in the AL-FS mode at 25 ± 1 °C. Therefore, these results are an important reference for the development of improved draw solutions in the FO system.

Author Contributions: Conceptualization, H.K.; Methodology, H.K.; validation, J.M. and H.K.; formal analysis, J.M.; investigation, J.M.; resources, H.K.; data curation, J.M.; writing—original draft preparation, J.M. and H.K.; writing—review and editing, J.M. and H.K.; visualization, J.M.; supervision, H.K.; project administration, H.K.; funding acquisition, H.K. All authors have read and agreed to the published version of the manuscript.

Funding: This research was funded by Dong-A University Research Fund (Grant Number: 2022).

Institutional Review Board Statement: Not applicable.

Informed Consent Statement: Not applicable.

Data Availability Statement: The data presented in this study are available on request from the corresponding author.

Acknowledgments: Financial supports by the Dong–A University Research Fund are gratefully acknowledged.

Conflicts of Interest: The authors declare no conflict of interest.

References

1. Dolan, F.; Lamontagne, J.; Link, R.; Hejazi, M.; Reed, P.; Edmonds, J. Evaluating the Economic Impact of Water Scarcity in a Changing World. *Nat. Commun.* **2021**, *12*, 1915. [CrossRef] [PubMed]
2. Mekonnen, M.M.; Hoekstra, A.Y. Four Billion People Facing Severe Water Scarcity. *Sci. Adv.* **2016**, *2*, e1500323. [CrossRef] [PubMed]
3. Elimelech, M.; Phillip, W.A. The Future of Seawater Desalination: Energy, Technology, and The Environment. *Science* **2011**, *333*, 712–717. [CrossRef] [PubMed]
4. Coday, B.D.; Yaffe, B.G.; Xu, P.; Cath, T.Y. Rejection of Trace Organic Compounds by Forward Osmosis Membranes: A Literature Review. *Environ. Sci. Technol.* **2014**, *48*, 3612–3624. [CrossRef] [PubMed]
5. Linares, R.V.; Li, Z.; Sarp, S.; Bucs, S.S.; Amy, G.; Vrouwenvelder, J.S. Forward Osmosis Niches in Seawater Desalination and Wastewater Reuse. *Water Res.* **2014**, *66*, 122–139. [CrossRef]
6. Wilf, M. Future of the Osmotic Processes. *Desalin. Water Treat.* **2010**, *15*, 292–298. [CrossRef]
7. Cath, T.Y.; Childress, A.E.; Elimelech, M. Forward Osmosis: Principles, Applications, and Recent Developments. *J. Membr. Sci.* **2006**, *281*, 70–87. [CrossRef]
8. Suwaileh, W.; Pathak, N.; Shon, H.; Hilal, N. Forward Osmosis Membranes and Processes: A Comprehensive Review of Research Trends and Future Outlook. *Desalination* **2020**, *485*, 114455. [CrossRef]
9. Kwan, S.E.; Bar-Zeev, E.; Elimelech, M. Biofouling in Forward Osmosis and Reverse Osmosis: Measurements and Mechanisms. *J. Membr. Sci.* **2015**, *493*, 703–708. [CrossRef]
10. Xie, M.; Lee, J.; Nghiem, L.D.; Elimelech, M. Role of Pressure in Organic Fouling in Forward Osmosis and Reverse Osmosis. *J. Membr. Sci.* **2015**, *493*, 748–754. [CrossRef]
11. Kim, Y.; Elimelech, M.; Shon, H.K.; Hong, S. Combined Organic and Colloidal Fouling in Forward Osmosis: Fouling Reversibility and the Role of Applied Pressure. *J. Membr. Sci.* **2014**, *460*, 206–212. [CrossRef]
12. Johnson, D.J.; Suwaileh, W.A.; Mohammed, A.W.; Hilal, N. Osmotic's Potential: An Overview of Draw Solutes for Forward Osmosis. *Desalination* **2018**, *434*, 100–120. [CrossRef]
13. Takahashi, T.; Akiya, K.; Niizeki, T.; Matsumoto, M.; Hoshina, T. Tunable Thermoresponsive UCST–Type Alkylimidazolium Ionic Liquids as a Draw Solution in the Forward Osmosis Process. *Colloids Surf. Physicochem. Eng. Asp.* **2022**, *639*, 128372. [CrossRef]
14. Chung, T.; Li, X.; Ong, R.C.; Ge, Q.; Wang, H.; Han, G. Emerging Forward Osmosis (FO) Technologies and Challenges Ahead for Clean Water and Clean Energy Applications. *Curr. Opin.* **2012**, *1*, 246–257. [CrossRef]
15. Phuntsho, S.; Shon, H.K.; Hong, S.; Lee, S.; Vigneswaran, S. A Novel Low Energy Fertilizer Driven Forward Osmosis Desalination for Direct Fertigation: Evaluating the Performance of Fertilizer Draw Solutions. *J. Membr. Sci.* **2011**, *375*, 172–181. [CrossRef]

16. Lutchmiah, K.; Lauber, L.; Roest, K.; Harmsen, D.J.; Post, J.W.; Rietveld, L.C.; van Lier, J.B.; Cornelissen, E.R. Zwitterions as Alternative Draw Solutions in Forward Osmosis for Application in Wastewater Reclamation. *J. Membr. Sci.* **2014**, *460*, 82–90. [CrossRef]
17. Nguyen, H.T.; Chen, S.; Nguyen, N.C.; Ngo, H.H.; Guo, W.; Li, C. Exploring an Innovative Surfactant and Phosphate-Based Draw Solution for Forward Osmosis Desalination. *J. Membr. Sci.* **2015**, *489*, 212–219. [CrossRef]
18. Ju, C.; Kang, H. Zwitterionic Polymers Showing Upper Critical Solution Temperature Behavior as Draw Solutes for Forward Osmosis. *RSC Adv.* **2017**, *7*, 56426–56432. [CrossRef]
19. Bai, Y.; Liang, Y.N.; Hu, X. An Eco-Friendly Approach for Heavy Metal Adsorbent Regeneration using CO_2-Responsive Molecular Octopus. *Chemosphere* **2017**, *185*, 1157–1163. [CrossRef]
20. Park, S.Y.; Ahn, H.; Chung, J.W.; Kwak, S. Magnetic Core-Hydrophilic Shell Nanosphere as Stability-Enhanced Draw Solute for Forward Osmosis (FO) Application. *Desalination* **2016**, *397*, 22–29. [CrossRef]
21. Kim, J.; Chung, J.; Kang, H.; Yu, Y.A.; Choi, W.J.; Kim, H.J.; Lee, J. Thermo-Responsive Copolymers with Ionic Group as Novel Draw Solutes for Forward Osmosis Processes. *Macromol. Res.* **2014**, *22*, 963–970. [CrossRef]
22. Wu, Y.; Liu, Y.; Chen, R.; Zhang, W.; Ge, Q. A pH-Responsive Supramolecular Draw Solute that Achieves High-Performance in Arsenic Removal Via Forward Osmosis. *Water Res.* **2019**, *165*, 114993. [CrossRef] [PubMed]
23. Riabtseva, A.; Ellis, S.N.; Champagne, P.; Jessop, P.G.; Cunningham, M.F. CO_2-Responsive Branched Polymers for Forward Osmosis Applications: The Effect of Branching on Draw Solute Properties. *Ind. Eng. Chem. Res.* **2021**, *60*, 9807–9816. [CrossRef]
24. Ellis, S.N.; Riabtseva, A.; Dykeman, R.R.; Hargreaves, S.; Robert, T.; Champagne, P.; Cunningham, M.F.; Jessop, P.G. Nitrogen Rich CO_2-Responsive Polymers as Forward Osmosis Draw Solutes. *Ind. Eng. Chem. Res.* **2019**, *58*, 22579–22586. [CrossRef]
25. Cui, H.; Zhang, H.; Yu, M.; Yang, F. Performance Evaluation of Electric-Responsive Hydrogels as Draw Agent in Forward Osmosis Desalination. *Desalination* **2018**, *426*, 118–126. [CrossRef]
26. MingáLing, M. Facile Synthesis of Thermosensitive Magnetic Nanoparticles as "smart" Draw Solutes in Forward Osmosis. *Chem. Commun.* **2011**, *47*, 10788–10790.
27. Fresnais, J.; Yan, M.; Courtois, J.; Bostelmann, T.; Bée, A.; Berret, J. Poly (Acrylic Acid)-Coated Iron Oxide Nanoparticles: Quantitative Evaluation of the Coating Properties and Applications for the Removal of a Pollutant Dye. *J. Colloid Interface Sci.* **2013**, *395*, 24–30. [CrossRef]
28. Zhao, Q.; Chen, N.; Zhao, D.; Lu, X. Thermoresponsive Magnetic Nanoparticles for Seawater Desalination. *ACS Appl. Mater. Interfaces* **2013**, *5*, 11453–11461. [CrossRef]
29. Cai, Y. A Critical Review on Draw Solutes Development for Forward Osmosis. *Desalination* **2016**, *391*, 16–29. [CrossRef]
30. Cai, Y.; Shen, W.; Wei, J.; Chong, T.H.; Wang, R.; Krantz, W.B.; Fane, A.G.; Hu, X. Energy-Efficient Desalination by Forward Osmosis using Responsive Ionic Liquid Draw Solutes. *Environ. Sci. Water Res. Technol.* **2015**, *1*, 341–347. [CrossRef]
31. Nakayama, D.; Mok, Y.; Noh, M.; Park, J.; Kang, S.; Lee, Y. Lower Critical Solution Temperature (LCST) Phase Separation of Glycol Ethers for Forward Osmotic Control. *Phys. Chem. Chem. Phys.* **2014**, *16*, 5319–5325. [CrossRef]
32. Li, D.; Zhang, X.; Yao, J.; Simon, G.P.; Wang, H. Stimuli-Responsive Polymer Hydrogels as a New Class of Draw Agent for Forward Osmosis Desalination. *Chem. Commun.* **2011**, *47*, 1710–1712. [CrossRef]
33. Shakeri, A.; Salehi, H.; Taghvay Nakhjiri, M.; Shakeri, E.; Khankeshipour, N.; Ghorbani, F. Carboxymethylcellulose-Quaternary Graphene Oxide Nanocomposite Polymer Hydrogel as a Biodegradable Draw Agent for Osmotic Water Treatment Process. *Cellulose* **2019**, *26*, 1841–1853. [CrossRef]
34. Kim, T.; Ju, C.; Park, C.; Kang, H. Polymer having Dicationic Structure in Dumbbell Shape for Forward Osmosis Process. *Polymers* **2019**, *11*, 571. [CrossRef]
35. Endo, T.; Sunada, K.; Sumida, H.; Kimura, Y. Origin of Low Melting Point of Ionic Liquids: Dominant Role of Entropy. *Chem. Sci.* **2022**, *13*, 7560–7565. [CrossRef]
36. Kokorin, A. *Ionic Liquids: Theory, Properties, New Approaches*; BoD—Books on Demand: Norderstedt, Germany, 2011.
37. Wilkes, J.S.; Zaworotko, M.J. Air and Water Stable 1-Ethyl-3-Methylimidazolium Based Ionic Liquids. *J. Chem. Soc. Chem. Commun.* **1992**, 965–967. [CrossRef]
38. Welton, T. Room-Temperature Ionic Liquids. Solvents for Synthesis and Catalysis. *Chem. Rev.* **1999**, *99*, 2071–2084. [CrossRef]
39. Plechkova, N.V.; Seddon, K.R. Applications of Ionic Liquids in the Chemical Industry. *Chem. Soc. Rev.* **2008**, *37*, 123–150. [CrossRef]
40. Wasserscheid, P.; Welton, T. *Ionic Liquids in Synthesis*; Wiley Online Library: Hoboken, NJ, USA, 2008.
41. Miao, W.; Chan, T.H. Ionic-Liquid-Supported Synthesis: A Novel Liquid-Phase Strategy for Organic Synthesis. *Acc. Chem. Res.* **2006**, *39*, 897–908. [CrossRef]
42. Cui, J.; Li, Y.; Chen, D.; Zhan, T.; Zhang, K. Ionic Liquid-based Stimuli-responsive Functional Materials. *Adv. Funct. Mater.* **2020**, *30*, 2005522. [CrossRef]
43. Abdullah, M.; Man, M.S.; Abdullah, S.B.; Saufi, S.M. Synthesis and Characterization of Thermo-Responsive Ionic Liquids (TRILs). *IOP Conf. Ser. Mater. Sci. Eng.* **2020**, *736*, 042027. [CrossRef]
44. Qiao, Y.; Ma, W.; Theyssen, N.; Chen, C.; Hou, Z. Temperature-Responsive Ionic Liquids: Fundamental Behaviors and Catalytic Applications. *Chem. Rev.* **2017**, *117*, 6881–6928. [CrossRef] [PubMed]

45. Zeweldi, H.G.; Bendoy, A.P.; Park, M.J.; Shon, H.K.; Kim, H.; Johnson, E.M.; Kim, H.; Lee, S.; Chung, W.; Nisola, G.M. Tetrabutylammonium 2, 4, 6-Trimethylbenzenesulfonate as an Effective and Regenerable Thermo-Responsive Ionic Liquid Drawing Agent in Forward Osmosis for Seawater Desalination. *Desalination* **2020**, *495*, 114635. [CrossRef]
46. Nyan, P.S.; Saufi, S.M.; Abdullah, S.B.; Seman, M.N.A.; Taib, M.M. Tetrabutylphosphonium Trifluoroacetate ([P_{4444}] CF_3COO) Thermoresponsive Ionic Liquid as a Draw Solution for Forward Osmosis Process. *Malays. J. Anal. Sci.* **2018**, *22*, 605–611.
47. Abdullah, M.A.M.; Seman, M.N.A.; Chik, S.M.S.T.; Abdullah, S.B. Factorial Design in Optimizing Parameters for Thermoresponsive Ionic Liquids as Draw Solution. *Process Saf. Environ. Prot.* **2022**, *161*, 34–49. [CrossRef]
48. Zeweldi, H.G.; Limjuco, L.A.; Bendoy, A.P.; Kim, H.; Park, M.J.; Shon, H.K.; Johnson, E.M.; Lee, H.; Chung, W.; Nisola, G.M. The Potential of Monocationic Imidazolium-, Phosphonium-, and Ammonium-Based Hydrophilic Ionic Liquids as Draw Solutes for Forward Osmosis. *Desalination* **2018**, *444*, 94–106. [CrossRef]
49. Zhong, Y.; Feng, X.; Chen, W.; Wang, X.; Huang, K.; Gnanou, Y.; Lai, Z. Using UCST Ionic Liquid as a Draw Solute in Forward Osmosis to Treat High-Salinity Water. *Environ. Sci. Technol.* **2016**, *50*, 1039–1045. [CrossRef]
50. Ju, C.; Park, C.; Kim, T.; Kang, S.; Kang, H. Thermo-Responsive Draw Solute for Forward Osmosis Process; Poly (Ionic Liquid) having Lower Critical Solution Temperature Characteristics. *RSC Adv.* **2019**, *9*, 29493–29501. [CrossRef]
51. Zhao, Y.; Tian, L.; Pei, Y.; Wang, H.; Wang, J. Effect of Anionic Structure on the LCST Phase Behavior of Phosphonium Ionic Liquids in Water. *Ind. Eng. Chem. Res.* **2018**, *57*, 12935–12941. [CrossRef]
52. Chen, L.; Zhao, C.; Duan, X.; Zhou, J.; Liu, M. Finely Tuning the Lower Critical Solution Temperature of Ionogels by Regulating the Polarity of Polymer Networks and Ionic Liquids. *CCS Chem.* **2022**, *4*, 1386–1396. [CrossRef]
53. Xu, W.; Gao, X.; Zheng, L.; Lu, F. Ionic-Liquid-Based Aqueous Two-Phase Systems Induced by Intra-and Intermolecular Hydrogen Bonds. *Molecules* **2022**, *27*, 5307. [CrossRef]
54. Pei, Y.; Cao, Y.; Huang, Y.; Song, X.; Wang, H.; Zhao, Y.; Wang, J. Tunable LCST-Type Phase Behavior of $[FeCl_4]^-$-Based Ionic Liquids in Water. *Sci. China Chem.* **2016**, *59*, 587–593. [CrossRef]
55. Kohno, Y.; Ohno, H. Ionic Liquid/Water Mixtures: From Hostility to Conciliation. *Chem. Commun.* **2012**, *48*, 7119–7130. [CrossRef]
56. Kohno, Y.; Ohno, H. Temperature-Responsive Ionic Liquid/Water Interfaces: Relation between Hydrophilicity of Ions and Dynamic Phase Change. *Phys. Chem. Chem. Phys.* **2012**, *14*, 5063–5070. [CrossRef]
57. Kohno, Y.; Saita, S.; Men, Y.; Yuan, J.; Ohno, H. Thermoresponsive Polyelectrolytes Derived from Ionic Liquids. *Polym. Chem.* **2015**, *6*, 2163–2178. [CrossRef]
58. Maton, C.; De Vos, N.; Stevens, C.V. Ionic Liquid Thermal Stabilities: Decomposition Mechanisms and Analysis Tools. *Chem. Soc. Rev.* **2013**, *42*, 5963–5977. [CrossRef]
59. Bradaric, C.J.; Downard, A.; Kennedy, C.; Robertson, A.J.; Zhou, Y. Industrial Preparation of Phosphonium Ionic Liquids. *Green Chem.* **2003**, *5*, 143–152. [CrossRef]
60. Stalpaert, M.; Cirujano, F.G.; De Vos, D.E. Tetrabutylphosphonium Bromide Catalyzed Dehydration of Diols to Dienes and its Application in the Biobased Production of Butadiene. *ACS Catal.* **2017**, *7*, 5802–5809. [CrossRef]
61. Khazalpour, S.; Yarie, M.; Kianpour, E.; Amani, A.; Asadabadi, S.; Seyf, J.Y.; Rezaeivala, M.; Azizian, S.; Zolfigol, M.A. Applications of Phosphonium-Based Ionic Liquids in Chemical Processes. *J. Iran. Chem. Soc.* **2020**, *17*, 1775–1917. [CrossRef]
62. Pena, C.A.; Soto, A.; Rodríguez, H. Tetrabutylphosphonium Acetate and its Eutectic Mixtures with Common-Cation Halides as Solvents for Carbon Dioxide Capture. *Chem. Eng. J.* **2021**, *409*, 128191. [CrossRef]
63. Mikkola, S.; Robciuc, A.; Lokajova, J.; Holding, A.J.; Lämmerhofer, M.; Kilpelainen, I.; Holopainen, J.M.; King, A.W.; Wiedmer, S.K. Impact of Amphiphilic Biomass-Dissolving Ionic Liquids on Biological Cells and Liposomes. *Environ. Sci. Technol.* **2015**, *49*, 1870–1878. [CrossRef] [PubMed]
64. Thomas, S.; Rayaroth, M.P.; Menacherry, S.P.M.; Aravind, U.K.; Aravindakumar, C.T. Sonochemical Degradation of Benzenesulfonic Acid in Aqueous Medium. *Chemosphere* **2020**, *252*, 126485. [CrossRef] [PubMed]
65. Masoudian, Z.; Salehi-Lisar, S.Y.; Norastehnia, A. Phytoremediation Potential of Azolla Filiculoides for Sodium Dodecyl Benzene Sulfonate (SDBS) Surfactant Considering some Physiological Responses, Effects of Operational Parameters and Biodegradation of Surfactant. *Environ. Sci. Pollut. Res.* **2020**, *27*, 20358–20369. [CrossRef] [PubMed]
66. Li, W.; Wu, P. Unusual Thermal Phase Transition Behavior of an Ionic Liquid and Poly (Ionic Liquid) in Water with significantly Different LCST and Dynamic Mechanism. *Polym. Chem.* **2014**, *5*, 5578–5590. [CrossRef]
67. Yuan, W.; Yang, X.; He, L.; Xue, Y.; Qin, S.; Tao, G. Viscosity, Conductivity, and Electrochemical Property of Dicyanamide Ionic Liquids. *Front. Chem.* **2018**, *6*, 59. [CrossRef]
68. Dukhin, A.; Parlia, S.; Somasundaran, P. Ion-Pair Conductivity Theory V: Critical Ion Size and Range of Ion-Pair Existence. *J. Electrochem. Soc.* **2018**, *165*, E784. [CrossRef]
69. Campbell, R.B.; Bower, C.A.; Richards, L.A. Change of Electrical Conductivity with Temperature and the Relation of Osmotic Pressure to Electrical Conductivity and Ion Concentration for Soil Extracts. *Soil Sci. Soc. Am. J.* **1949**, *13*, 66–69. [CrossRef]
70. Every, H.; Bishop, A.G.; Forsyth, M.; Macfarlane, D.R. Ion Diffusion in Molten Salt Mixtures. *Electrochim. Acta* **2000**, *45*, 1279–1284. [CrossRef]
71. Sha, M.; Dong, H.; Luo, F.; Tang, Z.; Zhu, G.; Wu, G. Dilute or Concentrated Electrolyte Solutions? Insight from Ionic Liquid/Water Electrolytes. *J. Phys. Chem. Lett.* **2015**, *6*, 3713–3720. [CrossRef]
72. Otero-Mato, J.M.; Montes-Campos, H.; Gómez-González, V.; Montoto, M.; Cabeza, O.; Kondrat, S.; Varela, L.M. Structure, Dynamics and Conductivities of Ionic Liquid-Alcohol Mixtures. *J. Mol. Liq.* **2022**, *355*, 118955. [CrossRef]

73. Chaban, V.V.; Voroshylova, I.V.; Kalugin, O.N.; Prezhdo, O.V. Acetonitrile Boosts Conductivity of Imidazolium Ionic Liquids. *J. Phys. Chem. B* **2012**, *116*, 7719–7727. [CrossRef]
74. Rilo, E.; Vila, J.; García-Garabal, S.; Varela, L.M.; Cabeza, O. Electrical Conductivity of Seven Binary Systems Containing 1-Ethyl 3-Methyl Imidazolium Alkyl Sulfate Ionic Liquids with Water or Ethanol at Four Temperatures. *J. Phys. Chem. B* **2013**, *117* 1411–1418. [CrossRef]
75. Chen, M.; White, B.T.; Kasprzak, C.R.; Long, T.E. Advances in Phosphonium-Based Ionic Liquids and Poly (Ionic Liquid) s as Conductive Materials. *Eur. Polym. J.* **2018**, *108*, 28–37. [CrossRef]
76. McCutcheon, J.R.; McGinnis, R.L.; Elimelech, M. A Novel Ammonia—Carbon Dioxide Forward (Direct) Osmosis Desalination Process. *Desalination* **2005**, *174*, 1–11. [CrossRef]
77. Van't Hoff, J. The Function of Osmotic Pressure in the Analogy between Solutions and Gases. *Proc. Phys. Soc. Lond.* **1887**, *9*, 307 [CrossRef]
78. Silberberg-Bouhnik, M.; Ramon, O.; Ladyzhinski, I.; Mizrahi, S.; Cohen, Y. Osmotic Deswelling of Weakly Charged Poly (Acrylic Acid) Solutions and Gels. *J. Polym. Sci. B Polym. Phys.* **1995**, *33*, 2269–2279. [CrossRef]
79. Bates, S.J. Osmotic Pressure and Concentration in Solutions of Electrolytes, and the Calculation of the Degree of Ionization. *J. Am Chem. Soc.* **1915**, *37*, 1421–1445. [CrossRef]
80. Nordness, O.; Brennecke, J.F. Ion Dissociation in Ionic Liquids and Ionic Liquid Solutions. *Chem. Rev.* **2020**, *120*, 12873–12902 [CrossRef]
81. Freire, M.G.; Santos, L.M.; Fernandes, A.M.; Coutinho, J.A.; Marrucho, I.M. An Overview of the Mutual Solubilities of Water-imidazolium-Based Ionic Liquids Systems. *Fluid Phase Equilib.* **2007**, *261*, 449–454. [CrossRef]
82. Khan, I.; Kurnia, K.A.; Mutelet, F.; Pinho, S.P.; Coutinho, J.A. Probing the Interactions between Ionic Liquids and Water Experimental and Quantum Chemical Approach. *J. Phys. Chem. B* **2014**, *118*, 1848–1860. [CrossRef]
83. Freire, M.G.; Carvalho, P.J.; Gardas, R.L.; Marrucho, I.M.; Santos, L.M.; Coutinho, J.A. Mutual Solubilities of Water and the [Cnmim][Tf2N] Hydrophobic Ionic Liquids. *J. Phys. Chem. B* **2008**, *112*, 1604–1610. [CrossRef] [PubMed]
84. Belchior, D.C.; Sintra, T.E.; Carvalho, P.J.; Soromenho, M.R.; Esperança, J.M.; Ventura, S.P.; Rogers, R.D.; Coutinho, J.A.; Freire M.G. Odd-Even Effect on the Formation of Aqueous Biphasic Systems Formed by 1-Alkyl-3-Methylimidazolium Chloride Ionic Liquids and Salts. *J. Chem. Phys.* **2018**, *148*, 193842. [CrossRef] [PubMed]
85. Ou, R.; Wang, Y.; Wang, H.; Xu, T. Thermo-Sensitive Polyelectrolytes as Draw Solutions in Forward Osmosis Process. *Desalination* **2013**, *318*, 48–55. [CrossRef]
86. Li, D.; Zhang, X.; Simon, G.P.; Wang, H. Forward Osmosis Desalination using Polymer Hydrogels as a Draw Agent: Influence of Draw Agent, Feed Solution and Membrane on Process Performance. *Water Res.* **2013**, *47*, 209–215. [CrossRef]
87. Zhao, C.; Ma, Z.; Zhu, X.X. Rational Design of Thermoresponsive Polymers in Aqueous Solutions: A Thermodynamics Map *Prog. Polym. Sci.* **2019**, *90*, 269–291. [CrossRef]
88. Pasparakis, G.; Tsitsilianis, C. LCST Polymers: Thermoresponsive Nanostructured Assemblies Towards Bioapplications. *Polymer* **2020**, *211*, 123146. [CrossRef]
89. Kang, H.; Suich, D.E.; Davies, J.F.; Wilson, A.D.; Urban, J.J.; Kostecki, R. Molecular Insight into the Lower Critical Solution Temperature Transition of Aqueous Alkyl Phosphonium Benzene Sulfonates. *Commun. Chem.* **2019**, *2*, 51. [CrossRef]
90. Wang, Z.; Elimelech, M.; Lin, S. Environmental Applications of Interfacial Materials with Special Wettability. *Environ. Sci. Technol.* **2016**, *50*, 2132–2150. [CrossRef]
91. Tiraferri, A.; Kang, Y.; Giannelis, E.P.; Elimelech, M. Highly Hydrophilic Thin-Film Composite Forward Osmosis Membranes Functionalized with Surface-Tailored Nanoparticles. *ACS Appl. Mater. Interfaces* **2012**, *4*, 5044–5053. [CrossRef]
92. Gao, K.; Kearney, L.T.; Wang, R.; Howarter, J.A. Enhanced Wettability and Transport Control of Ultrafiltration and Reverse Osmosis Membranes with Grafted Polyelectrolytes. *ACS Appl. Mater. Interfaces* **2015**, *7*, 24839–24847. [CrossRef]
93. Ismail, M.F.; Islam, M.A.; Khorshidi, B.; Tehrani-Bagha, A.; Sadrzadeh, M. Surface Characterization of Thin-Film Composite Membranes using Contact Angle Technique: Review of Quantification Strategies and Applications. *Adv. Colloid Interface Sci.* **2022**, *299*, 102524. [CrossRef]
94. Ferby, M.; Zou, S.; He, Z. Reduction of reverse solute flux induced solute buildup in the feed solution of forward osmosis. *Environ. Sci. Water Res. Technol.* **2020**, *6*, 423–435. [CrossRef]
95. Gray, G.T.; McCutcheon, J.R.; Elimelech, M. Internal Concentration Polarization in Forward Osmosis: Role of Membrane Orientation. *Desalination* **2006**, *197*, 1–8. [CrossRef]
96. McCutcheon, J.R.; Elimelech, M. Modeling Water Flux in Forward Osmosis: Implications for Improved Membrane Design. *AIChE J.* **2007**, *53*, 1736–1744. [CrossRef]
97. McCutcheon, J.R.; Elimelech, M. Influence of Concentrative and Dilutive Internal Concentration Polarization on Flux Behavior in Forward Osmosis. *J. Membr. Sci.* **2006**, *284*, 237–247. [CrossRef]

membranes

MDPI

Article

Microstructure and Hydrogen Permeability of Nb-Ni-Ti-Zr-Co High Entropy Alloys

Egor Kashkarov [1,*], Dmitriy Krotkevich [1], Maxim Koptsev [1], Sergei Ognev [1], Leonid Svyatkin [1], Nahum Travitzky [1,2] and Andrey Lider [1]

[1] School of Nuclear Science and Engineering, National Research Tomsk Polytechnic University, 634050 Tomsk, Russia
[2] Department of Materials Science, Glass and Ceramics, Friedrich-Alexander-Universität Erlangen-Nürnberg, 91054 Erlangen, Germany
* Correspondence: ebk@tpu.ru; Tel.: +7-3822-70-17-77 (ext. 1562)

Abstract: Hydrogen separation membranes are one of the most promising technologies for hydrogen purification. The development of high-entropy alloys (HEAs) for hydrogen separation membranes is driven by a "cocktail effect" of elements with different hydrogen affinities to prevent hydride formation and retain high permeability due to the single-phase BCC structure. In this paper, equimolar and non-equimolar Nb-Ni-Ti-Zr-Co high entropy alloys were fabricated by arc melting. The microstructure and phase composition of the alloys were analyzed by scanning electron microscopy and X-ray diffraction, respectively. The hydrogen permeation experiments were performed at 300–500 °C and a hydrogen pressure of 4 bar. In order to estimate the effect of composition and lattice structure on hydrogen location and diffusivity in Nb-Ni-Ti-Zr-Co alloy, ab initio calculations of hydrogen binding energy were performed using virtual crystal approximation. It was found that Nb-enriched and near equimolar BCC phases were formed in $Nb_{20}Ni_{20}Ti_{20}Zr_{20}Co_{20}$ HEA while Nb-enriched BCC and B2-Ni(Ti, Zr) were formed in $Nb_{40}Ni_{25}Ti_{18}Zr_{12}Co_5$ alloy. Hydrogen permeability tests showed that $Nb_{20}Ni_{20}Ti_{20}Zr_{20}Co_{20}$ HEA shows lower activation energy and higher permeability at lower temperatures as well as higher resistance to hydrogen embrittlement compared to $Nb_{40}Ni_{25}Ti_{18}Zr_{12}Co_5$ alloy. The effect of composition, microstructure and hydrogen binding energies on permeability of the fabricated alloys was discussed.

Keywords: high-entropy alloys; membranes; microstructure; ab initio calculation; hydrogen permeability

1. Introduction

Membrane technology is one of the most efficient approaches for hydrogen separation and purification due to low operating costs, high productivity and low power consumption [1,2]. Metal membranes have the highest selectivity for hydrogen separation. Thin Pd foils have a separation factor >1000 [1]. Palladium-based (Pd-based) membranes with a face centered cubic (fcc) lattice structure have the ability to dissociate and dissolve molecular hydrogen and thus exhibit excellent H_2 permeability properties [3]. However, the high cost of Pd motivates researchers to develop composite membranes or alternative effective membrane materials [4].

The V metals group (V, Nb, Ta, etc.), with a more open body-centered cubic (BCC) lattice, are more permeable to hydrogen and are promising candidate materials for hydrogen separation membranes [5,6]. However, in most cases the single-element membranes of the V group suffer from hydrogen embrittlement (HE) and gas poisoning at high temperature, which does not allow their wide use for hydrogen separation in industry [7]. Thus, multicomponent alloys based on V group elements with a BCC structure, which are more resistant to embrittlement, are being developed [8,9]. It is well known that the composition of alloys significantly affects its hydrogen permeability and resistance to HE [10].

It was shown that alloying of Nb-based alloys with Ti and Ni can lead to a reduction of hydrogen embrittlement and permeability [11,12]. In addition to the formation of BCC phases, interdendritic TiNi compounds inhibit hydride formation. Ni stabilizes the alloy with reduced H absorption. The presence of Ti in Nb-based alloys increases the thermal stability and hydrogen absorption [9,13]. The addition of Zr in Ni-Nb-Zr-Co alloy increases hydrogen sorption and increases permeability; however, it reduces the resistance to HE. The addition of Co is effective in suppressing hydrogen embrittlement of Ni-Nb-Zr-Co alloy [14]. Tang et al. investigated multicomponent $Nb_{40}Ti_{18}Zr_{12}Ni_{25}M_5$ (M = Al, Co, Cu, Pd) and showed that the $Nb_{40}Ti_{18}Zr_{12}Ni_{25}Co_5$ alloy had the highest permeability of 3.8×10^{-8} ($molH_2m^{-1}s^{-1}Pa^{-0.5}$), higher than that for $Pd_{0.23}Ag_{0.77}$ alloy, with a permeability of 1.6×10^{-8} ($molH_2m^{-1}s^{-1}Pa^{-0.5}$), while maintaining stable performance [15].

Multicomponent alloys with more than four elements in near equimolar composition, also known as high-entropy alloys (HEAs), can be considered as membrane materials. The "cocktail effect" [16] of elements with different hydrogen affinities can provide low hydrogen embrittlement and high permeability due to the single phase BCC crystal structure. It was reported that high-entropy alloys have high mechanical strength, which is maintained over a wide temperature range, phase stability and low degradation rate [17]. Han et al. demonstrated high phase stability and mechanical properties for $Nb_{20}Ni_{20}Ti_{20}Co_{20}Zr_{20}$ under compressive tests up to 800 °C [18]. However, there are no experimental works on the hydrogen permeability of this HEA. Therefore, in the present work, based on the positive effect of alloying elements we selected the Nb-Ni-Ti-Zr-Co system to fabricate equimolar and non-equimolar high-entropy alloys and investigated its microstructure and hydrogen permeability.

2. Materials and Methods

2.1. Material Preparation

Metal powders of Nb, Ni, Ti, Zr, Co (>99.5% purity) were mixed in a ball mill (AGO-2: Novits, Novosibirsk, Russia) to obtain homogeneous distribution of elemental powders. The powders were mixed with the rotation speed of 170 rpm for 1 h. Two series of powders with equimolar composition ($Nb_{20}Ni_{20}Ti_{20}Zr_{20}Co_{20}$) and non-equimolar composition ($Nb_{40}Ni_{25}Ti_{18}Zr_{12}Co_5$) were prepared. The mixed powders were pressed by a cold static uniaxial method at the pressure of 90 MPa in a closed mold. The alloy samples were fabricated by arc melting of the pressed powders on a water-cooled copper crucible in an argon atmosphere. Then, the samples were annealed in vacuum at 800 °C for 10 h. The resulting ingots were cut into disks of 10 mm in diameter and 0.7 mm in thickness. Then, the samples were ground using SiC paper with a grain size up to 5 μm. Finally, the samples were annealed in vacuum at 800 °C for 6 h.

2.2. Characterization

The phase composition and crystalline structure of the fabricated alloys were analyzed by X-ray diffraction (XRD) using an XRD-7000S diffractometer (Shimadzu, Kyoto, Japan) with CuK_α radiation (1.5410 Å wavelength) at 40 kV and 30 mA in the Bragg-Brentano geometry. The phases were identified using the PDF4+ 2021 database and SIeve software (ICDD, Newtown, PA, USA). The phase composition and lattice parameters were calculated by Rietveld refinement. The microstructure and elemental composition of the samples were analyzed by scanning electron microscopy (SEM) using Vega 3 (Tescan, Brno, Czech Republic) equipped with an energy dispersive X-ray spectroscopy (EDX) attachment.

2.3. Hydrogen Permeation Tests

For the hydrogen permeation test, the samples were placed in the test cell with a copper seal gasket (Figure 1). The surface of the samples was covered by thin Pd film using magnetron sputtering. The active surface area of the membrane samples was 1.9×10^{-5} m^2. Leak testing was carried out before and after the experiment. Hydrogen was supplied from one side of the membrane at room temperature. From the side of the gas outlet, the hydrogen signal was analyzed by a mass spectrometer. The experiment was started after the leakage test was valid. The chamber was evacuated to the residual pressure of 10^{-3} Pa from both sides of the sample. The sample was heated to 300–500 °C with a heating rate of 6 °C/min. After reaching the heating temperature, hydrogen gas (4 bar) was supplied to the one side of the sample.

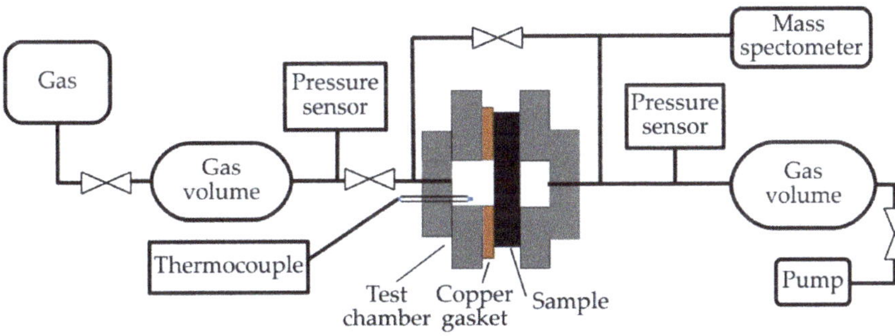

Figure 1. Schematic diagram of the test cell for hydrogen permeation experiments.

The concentration of hydrogen permeated through the sample was measured by the change in pressure in known volumes:

$$n(mol) = \frac{P \times V}{R \times T}, \tag{1}$$

where P—pressure (Pa), V—volume (m^3), R—gas constant, T—temperature (K).

The flux (Φ) was calculated from the difference in the amount of gas passed through the area of the sample in a period of time:

$$\Phi = \frac{n1 - n2}{t \times S}, \tag{2}$$

where $n1$—initial concentration of gas (mol), $n2$—final concentration of gas (mol), t—time (s), S—active surface area of the sample (m^2).

Hydrogen permeability was calculated according to the following expression [19]:

$$j = \Phi \times \frac{d}{\Delta P^{0.5}}, \tag{3}$$

where d—sample thickness, ΔP—pressure difference.

2.4. *Ab Initio Calculations*

The binding energy for a hydrogen atom in Nb-Ni-Ti-Zr-Co high entropy alloys was theoretically investigated using first principles. All calculations were performed within the framework of the density functional theory using the optimized norm-preserving Vanderbilt pseudopotential [20], realized in the ABINIT software package [21,22] within virtual crystal approximation (VCA) [23,24]. To describe the exchange and correlation effects, the generalized gradient approximation in the form of Perdew, Burke, and Ernzerhof was used [25]. The lattices were completely relaxed. The relaxation was considered complete when the forces acting on the atoms were less than 25 meV/Å. To carry out the structural optimization and relaxation of the Nb-Ni-Ti-Zr-Co system for discussion, a cell with 2 atoms with mixing potentials and 1 H atom was adopted, and the k meshes were chosen to be 10 × 10 × 10 for BCC structures. The cutoff energy for the plane-wave basis was set at 820 eV.

The binding energy of hydrogen atom in HEAs was calculated according to the following expression:

$$E_b = E(A_2) + ½E(H_2) - E(A_2H), \quad (4)$$

where $E(A_2)$ is the energy of pure alloy in the presence of two mixing atoms in the cell; $E(H_2)$ is the energy of the hydrogen molecule; $E(A_2H)$ is the energy of the structure when there is a hydrogen atom in it.

3. Results

3.1. Microstructure and Phase Composition

The XRD patterns of $Nb_{20}Ni_{20}Ti_{20}Zr_{20}Co_{20}$ and $Nb_{40}Ni_{25}Ti_{18}Zr_{12}Co_5$ high entropy alloys after annealing at 800 °C for 10 h are presented in Figure 2. According to XRD measurements, equimolar $Nb_{20}Ni_{20}Ti_{20}Zr_{20}Co_{20}$ HEA consists of two BCC phases with lattice constants of 3.306 Å and 3.093 Å (Figure 2a). For non-equimolar $Nb_{40}Ni_{25}Ti_{18}Zr_{12}Co_5$ alloy, an Nb-rich BCC phase with the lattice constant of 3.306 Å and a B2-Ni(Ti, Zr) phase with the lattice constant of 3.060 Å were observed (Figure 2b). Additionally, weakly pronounced peaks of B2-CoZr compound were detected for the non-equimolar alloy.

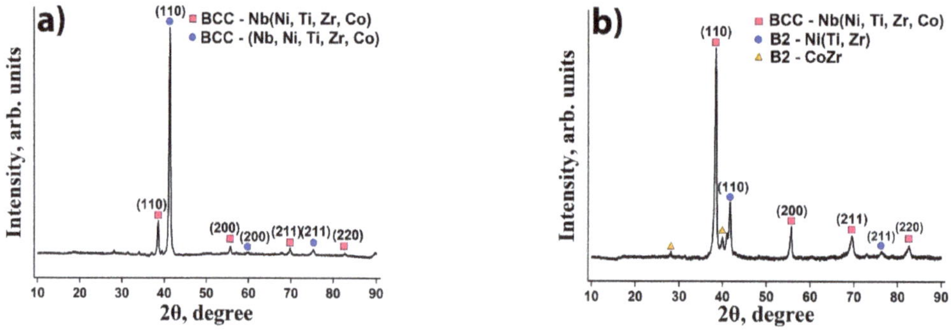

Figure 2. XRD patterns of $Nb_{20}Ni_{20}Ti_{20}Zr_{20}Co_{20}$ (**a**) and $Nb_{40}Ni_{25}Ti_{18}Zr_{12}Co_5$ (**b**) high entropy alloys annealed at 800 °C for 10 h.

Figure 3a shows the SEM image and corresponding EDS maps of the equimolar $Nb_{20}Ni_{20}Ti_{20}Zr_{20}Co_{20}$ alloy. Two major regions with different contrast can be seen: a light-grey contrasted Nb-rich BCC phase (region 1) with a globular and dendritic structure and a dark-grey contrasted—(Nb, Ni, Ti, Zr, Co) BCC phase (region 2), which has elemental content close to equimolar composition (Table 1).

Table 1. Elemental concentration in different regions as depicted in Figure 3.

Sample	Region	Concentration, at. %					Possible Phases
		Nb	Ni	Ti	Zr	Co	
$Nb_{20}Ni_{20}Ti_{20}Zr_{20}Co_{20}$	1	74	6	9	5	6	BCC-Nb(Ni, Ti, Zr, Co)
	2	15	20	16	29	19	BCC-(Nb, Ni, Ti, Zr, Co)
$Nb_{40}Ni_{25}Ti_{18}Zr_{12}Co_5$	3	88.8	3.7	6.5	-	0.9	BCC-Nb(Ni, Ti, Co)
	4	7.7	41.6	26.3	16.6	7.6	B2-Ni(Ti, Zr), B2-CoZr
	5	31	36	17.7	8.5	6.5	BCC-Nb(Ni, Ti, Zr, Co), B2-Ni(Ti, Zr)

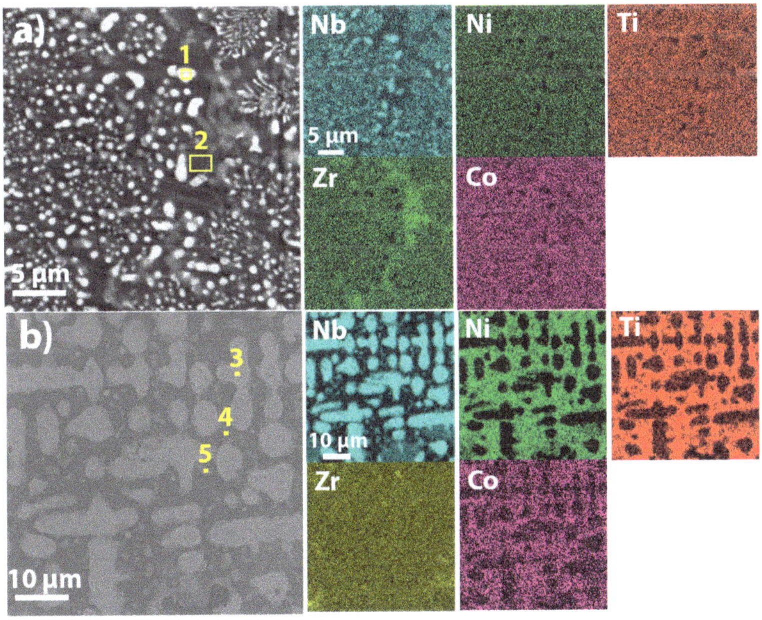

Figure 3. SEM images of $Nb_{20}Ni_{20}Ti_{20}Zr_{20}Co_{20}$ (**a**) and $Nb_{40}Ni_{25}Ti_{18}Zr_{12}Co_5$ (**b**) HEAs and corresponding EDS maps.

The microstructure of non-equimolar $Nb_{40}Ni_{25}Ti_{18}Zr_{12}Co_5$ (Figure 3b) high entropy alloy is represented by a light-grey contrasted Nb-rich BCC phase with dendritic structure (region 3) embedded in dark-grey contrasted matrix of B2-Ni(Ti, Zr) and B2-CoZr phases (region 4). Similar morphology and phases were observed by Tang et al. for $Nb_{40}Ti_{18}Zr_{12}Ni_{30}$ alloy [15]. Additionally, light-grey globular inclusions with high concentration of Nb (31 at.%) were detected in region 5. Since this also has high Ni content, it is assumed that this region corresponds to an Nb-enriched BCC-phase surrounded by a B2-Ni(Ti, Zr) phase.

3.2. Hydrogen Permeability

Figure 4 shows permeability curves for the HEA samples. The permeability of $Nb_{40}Ni_{25}Ti_{18}Zr_{12}Co_5$ obtained by Tang et al. was added to Figure 4 for comparison. To analyze the temperature effect, Arrhenius plots were constructed based on the data of linear approximation of the permeability rate on inverse temperature. The permeability of the equimolar alloy is higher at lower temperatures (300–350 °C), but lower at higher temperatures compared to the non-equimolar alloy. Approximation of the permeability by the Arrhenius equation for HEA alloys showed that the activation energy of permeability for $Nb_{20}Ni_{20}Ti_{20}Co_{20}Zr_{20}$ is 30 kJ/mol with pre-exponential factor 1.8×10^{-6}, and for $Nb_{40}Ni_{25}Ti_{18}Zr_{12}Co_5$ the activation energy is 38 kJ/mol with pre-exponential factor 1×10^{-5}. It can be seen that the activation energy of the equimolar high-entropy alloy is significantly lower than that for the non-equimolar $Nb_{40}Ti_{18}Zr_{12}$ $Ni_{25}Co_5$ alloy. According to permeability tests, it was revealed that equimolar HEA is less susceptible to hydrogen embrittlement. The $Nb_{40}Ni_{25}Ti_{18}Zr_{12}Co_5$ alloy failed (crashed) after the 12 h permeation experiment. Under the same conditions, the equimolar HEA worked for more than 26 h without loss of performance.

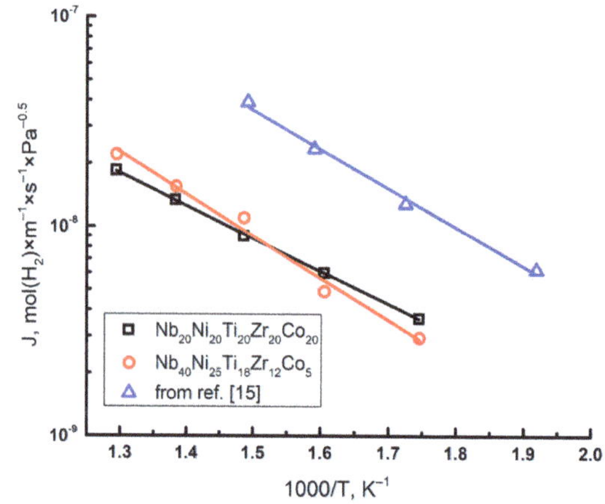

Figure 4. Hydrogen permeability of HEAs depending on temperature.

3.3. Lattice Constants and Hydrogen Binding Energy in HEAs

Table 2 summarizes the ab initio calculation results for lattice parameters and hydrogen binding energies in different phases of the HEAs. The overall dependence of the lattice constant changes depending on the elemental composition is in good correlation with the experimental results. The best agreement between the calculated results and experimental lattice parameters was observed for high niobium concentrations in the BCC lattice. However, in all cases the calculated lattice constant is underestimated, which can be caused by temperature effect, lattice strains in as-fabricated HEAs and features of the VCA.

Table 2. Lattice parameters and binding energies, *Eb*, for the hydrogen atom.

Alloy	Phase	Lattice Constant, Å				E_b, eV
		Exp	*a*, Calc	*b*, Calc	*c*, Calc	
$Nb_{74}Ni_6Ti_9Zr_5Co_6$	pure	3.306	3.192	3.192	3.192	-
	H in T sites		3.275	3.293	3.293	0.760
	H in O sites		3.137	3.137	3.586	0.441
$Nb_{15}Ni_{20}Ti_{16}Zr_{29}Co_{19}$	pure	3.090	2.979	2.979	2.979	-
	H in O sites		2.727	2.727	3.863	0.248
$Nb_{88.8}Ni_{3.7}Ti_{6.5}Co_{0.9}$	pure	3.306	3.251	3.251	3.251	-
	H in T sites		3.331	3.354	3.354	1.037
	H in O sites		3.196	3.196	3.641	0.777
$Nb_{7.7}Ni_{41.6}Ti_{26.32}Zr_{16.6}Co_{7.59}$	pure	3.060	2.855	2.855	2.855	-
	H in O sites		2.609	2.609	3.717	0.686

The difference in the changes in lattice parameters depending on the location of hydrogen atoms in BCC structure was found. The hydrogen atom in octahedral (O) sites leads to a lattice distortion along *c* axis at the octahedral coordination and all lattice expansion at the tetrahedral coordination in both BCC-Nb(Ni,Ti,Zr,Co) and BCC-(Nb,Ni,Ti,Zr,Co) phases.

The highest binding energy for hydrogen atoms was found at the high Nb concentration BCC lattice, which probably implies higher absorption of the hydrogen in regions with BCC-Nb(Ni,Ti,Zr,Co) phase. The decrease in Nb concentration leads to the decrease in the hydrogen binding energy in both T and O sites. In the $Nb_{15}Ni_{20}Ti_{16}Zr_{29}Co_{19}$ and

$Nb_{7.7}Ni_{41.6}Ti_{26.32}Zr_{16.6}Co_{7.59}$ systems, the location of the hydrogen atom in tetrahedral sites leads to system instability—a hydrogen atom leaves these sites. The lowest binding energy of the hydrogen atoms in the octahedral sites and the unstable tetrahedral configuration of hydrogen atoms can lead to the high hydrogen diffusivity in the BCC-(Nb,Ni,Ti,Zr,Co) lattice compared to Nb-rich BCC phase.

4. Discussion

The synthesized equimolar and non-equimolar HEAs have high hydrogen permeability, which depends on their microstructure and phase composition. The equimolar alloy has lower activation energy, and correspondingly higher permeability at lower temperatures (300–350 °C). This alloy is also more resistant to hydrogen embrittlement than the non-equimolar one containing a higher concentration of niobium. In the case of equimolar $Nb_{20}Ni_{20}Ti_{20}Zr_{20}Co_{20}$ HEA, a fine-dispersed structure with Nb-rich BCC-Nb(Ni,Ti,Zr,Co) and BCC-(Nb,Ni,Ti,Zr,Co) phases was obtained. Several studies have observed higher permeability of the multicomponent alloys with fine-dispersed structure, for example, for Nb–Ti–Co and Nb–Ni–Ti alloys [26–28]. Hashi et al. [26] and Yan et al. [27] also observed high resistance to hydrogen embrittlement for the fine grained alloy compared to coarse grained. Thus, the homogeneous fine microstructure of $Nb_{20}Ni_{20}Ti_{20}Zr_{20}Co_{20}$ HEA can have a positive effect both on the permeability of the alloy due to the large number of phase boundaries and on hydrogen embrittlement resistance. It is known that permeability depends on many factors, but the most important are solubility and diffusivity within the crystal lattice [29]. Niobium and its alloys are characterized by high hydrogen permeability due to high solubility limits and the relatively high diffusion mobility of hydrogen [30]. Based on ab initio calculations, hydrogen prefers to occupy tetrahedral sites in the Nb-based BCC lattice, which also agrees well with the results of other works [31]. The Nb-based crystal lattice also has a larger volume due to the larger lattice parameter of 3.306 A for the Nb-based BCC-Nb(Ni, Ti, Zr, Co) compared to the equimolar BCC-(Nb, Ni, Ti, Zr, Co) with the lattice parameter of 3.06 A. This provides higher solubility of the Nb-based BCC lattice in the HEAs. In turn, the low binding energy of hydrogen in the interstitial sites of the equimolar alloy can provide better diffusion mobility of hydrogen in the lattice but lower hydrogen absorption and hydrogen embrittlement. We believe that the competition between solubility and diffusivity can explain the observed difference in the permeability of HEAs. Based on the XRD data, the main phase in the equimolar alloy is BCC-(Nb, Ni, Ti, Zr, Co), while in the non-equimolar is BCC-Nb(Ni, Ti, Co). It should also be noted that the presence of secondary phases B2-Ni(Ti, Zr) and B2-CoZr in the $Nb_{40}Ni_{25}Ti_{18}Zr_{12}Co_{5}$ alloy can negatively affect its hydrogen permeability [26]. Thus, it is assumed that the high permeability of the $Nb_{40}Ni_{25}Ti_{18}Zr_{12}Co_{5}$ alloy at higher temperatures (400–500 °C) is provided mainly by the more permeable Nb-based BCC lattice with higher hydrogen solubility. While for the equimolar alloy, the high diffusivity of hydrogen through the BCC-(Nb, Ni, Ti, Zr, Co) along with the fine-dispersed structure provide lower activation energy and higher permeability at lower temperatures. In addition, it should be noted that the distortions of the crystal lattice in HEAs due to the difference in the atomic sizes in the structure make an additional contribution to the increase in hydrogen solubility. This can cause the difference in the behavior of permeability depending on the temperature for pure niobium and Nb-based multicomponent BCC structures.

5. Conclusions

High-entropy alloys with equimolar and non-equimolar compositions were synthesized by arc melting. Their microstructure, phase composition, hydrogen binding energy in interstitial sites as well as hydrogen permeability at 300–500 °C were studied. The following conclusions was made:

1. A fine-grained microstructure with BCC-(Nb, Ni, Ti, Zr, Co) and BCC-Nb(Ni, Ti, Zr, Co) lattices is formed in the equimolar $Nb_{20}Ni_{20}Ti_{20}Zr_{20}Co_{20}$ alloy, while coarser

dendritic microstructure with Nb-enriched BCC-Nb(Ni, Ti, Co), B2-Ni(Ti, Zr) and B2-CoZr phases is presented in the $Nb_{40}Ni_{25}Ti_{18}Zr_{12}Co_{5}$ alloy.

2. The lattice parameter and hydrogen binding energy in the interstitial sites are higher for the Nb-enriched BCC lattice than that for the BCC lattice with near equimolar composition, which indicates higher hydrogen solubility in the Nb-enriched phase. The hydrogen binding energy decreases in both tetrahedral and octahedral sites in the BCC-(Nb, Ni, Ti, Zr, Co) lattice, which apparently causes higher diffusivity of hydrogen.
3. The $Nb_{20}Ni_{20}Ti_{20}Zr_{20}Co_{20}$ alloy shows lower activation energy and higher permeability at temperatures of 300–350 °C as well as higher resistance to hydrogen embrittlement.

The obtained results give prospects for the development of HEA-based metallic membranes for hydrogen separation and purification.

Author Contributions: Conceptualization, E.K. and M.K.; Data curation, E.K.; Funding acquisition, E.K.; Formal analysis, S.O., L.S. and D.K.; Investigation, M.K., S.O. and E.K.; Methodology, M.K. and L.S.; Visualization, D.K.; Supervision, E.K.; Writing—original draft, M.K., D.K. and S.O.; Writing—review and editing, E.K., L.S., N.T. and A.L.; Project administration A.L, N.T. and E.K. All authors have read and agreed to the published version of the manuscript.

Funding: The research was funded by the Governmental Program, project No. FSWW-2021-0017.

Institutional Review Board Statement: Not applicable.

Data Availability Statement: Data are contained within the article.

Conflicts of Interest: The authors declare no conflict of interest.

References

1. Ockwig, N.W.; Nenoff, T.M. Membranes for Hydrogen Separation. *Chem. Rev.* **2007**, *107*, 4078–4110. [CrossRef]
2. Dolan, M.D. Non-Pd BCC Alloy Membranes for Industrial Hydrogen Separation. *J. Membr. Sci.* **2010**, *362*, 12–28. [CrossRef]
3. Conde, J.J.; Maroño, M.; Sánchez-Hervás, J.M. Pd-Based Membranes for Hydrogen Separation: Review of Alloying Elements and Their Influence on Membrane Properties. *Sep. Purif. Rev.* **2017**, *46*, 152–177. [CrossRef]
4. Lu, G.Q.; Diniz da Costa, J.C.; Duke, M.; Giessler, S.; Socolow, R.; Williams, R.H.; Kreutz, T. Inorganic Membranes for Hydrogen Production and Purification: A Critical Review and Perspective. *J. Colloid Interface Sci.* **2007**, *314*, 589–603. [CrossRef] [PubMed]
5. Li, F.; Zhong, B.; Xiao, H.; Ye, X.; Lu, L.; Guan, W.; Zhang, Y.; Wang, X.; Chen, C. Effect of Degassing Treatment on the Deuterium Permeability of Pd-Nb-Pd Composite Membranes during Deuterium Permeation. *Sep. Purif. Technol.* **2018**, *190*, 136–142 [CrossRef]
6. Phair, J.W.; Donelson, R. Developments and Design of Novel (Non-Palladium-Based) Metal Membranes for Hydrogen Separation *Ind. Eng. Chem. Res.* **2006**, *45*, 5657–5674. [CrossRef]
7. Yan, E.; Min, R.N.; Zhao, P.; Misra, R.D.K.; Huang, P.R.; Zou, Y.J.; Chu, H.L.; Zhang, H.Z.; Xu, F.; Sun, L.X. Design of Nb-Based Multi-Phase Alloy Membranes for High Hydrogen Permeability and Suppressed Hydrogen Embrittlement. *J. Membr. Sci.* **2020**, *595*, 117531. [CrossRef]
8. Nambu, T.; Shimizu, K.; Matsumoto, Y.; Rong, R.; Watanabe, N.; Yukawa, H.; Morinaga, M.; Yasuda, I. Enhanced Hydrogen Embrittlement of Pd-Coated Niobium Metal Membrane Detected by in Situ Small Punch Test under Hydrogen Permeation. *J. Alloys Compd.* **2007**, *446–447*, 588–592. [CrossRef]
9. Magnone, E.; Jeon, S.I.; Park, J.H.; Fleury, E. Relationship between Microstructure and Hydrogen Permeation Properties in the Multiphase Ni21Ti23Nb56 Alloy Membranes. *J. Membr. Sci.* **2011**, *384*, 136–141. [CrossRef]
10. Wang, N.J.; Liu, Y. Microstructure of Nb-Ta-V Based Hydrogen Permeation High Entropy Alloys. *Mater. Sci. Forum* **2016**, *849*, 3–7 [CrossRef]
11. Kishida, K.; Yamaguchi, Y.; Tanaka, K.; Inui, H.; Tokui, S.; Ishikawa, K.; Aoki, K. Microstructures and Hydrogen Permeability of Directionally Solidified Nb–Ni–Ti Alloys with the Nb–NiTi Eutectic Microstructure. *Intermetallics* **2008**, *16*, 88–95. [CrossRef]
12. Jiang, P.; Huang, H.; Sun, B.; Song, G.; Wu, W.; Wang, Z.; Zhang, Y. Microstructure, Mechanical Properties and Hydrogen Permeability of Multiphase V-Ti-Ni Alloy Membranes. *Mater. Today Commun.* **2020**, *24*, 101112. [CrossRef]
13. Belyakova, R.M.; Kurbanova, E.D.; Sidorov, N.I.; Polukhin, V.A. Nb–Ni- and V–Ni-Based Membranes for High-Purity Hydrogen Production. *Russ. Metall. Met.* **2022**, *2022*, 851–860. [CrossRef]
14. Shimpo, Y.; Yamaura, S.-I.; Nishida, M.; Kimura, H.; Inoue, A. Development of Melt-Spun Ni–Nb–Zr–Co Amorphous Alloy for High-Performance Hydrogen Separating Membrane. *J. Membr. Sci.* **2006**, *286*, 170–173. [CrossRef]
15. Tang, H.X.; Ishikawa, K.; Aoki, K. Effect of Elements Addition on Hydrogen Permeability and Ductility of Nb40Ti18Zr12Ni30 Alloy. *J. Alloys Compd.* **2008**, *461*, 263–266. [CrossRef]

16. Divinski, S.V.; Lukianova, O.A.; Wilde, G.; Dash, A.; Esakkiraja, N.; Paul, A. High-Entropy Alloys: Diffusion. In *Encyclopedia of Materials: Metals and Alloys*; Caballero, F.G., Ed.; Elsevier: Oxford, UK, 2022; pp. 402–416. ISBN 978-0-12-819733-2.
17. Cheng, C.-Y.; Yang, Y.-C.; Zhong, Y.-Z.; Chen, Y.-Y.; Hsu, T.; Yeh, J.-W. Physical Metallurgy of Concentrated Solid Solutions from Low-Entropy to High-Entropy Alloys. *Curr. Opin. Solid State Mater. Sci.* **2017**, *21*, 299–311. [CrossRef]
18. Han, Z.; Liu, X.; Zhao, S.; Shao, Y.; Li, J.; Yao, K. Microstructure, Phase Stability and Mechanical Properties of Nb–Ni–Ti–Co–Zr and Nb–Ni–Ti–Co–Zr–Hf High Entropy Alloys. *Prog. Nat. Sci. Mater. Int.* **2015**, *25*, 365–369. [CrossRef]
19. Suzuki, A.; Yukawa, H. A Review for Consistent Analysis of Hydrogen Permeability through Dense Metallic Membranes. *Membranes* **2020**, *10*, 120. [CrossRef] [PubMed]
20. Hamann, D.R. Optimized Norm-Conserving Vanderbilt Pseudopotentials. *Phys. Rev. B* **2013**, *88*, 085117. [CrossRef]
21. Gonze, X.; Amadon, B.; Antonius, G.; Arnardi, F.; Baguet, L.; Beuken, J.-M.; Bieder, J.; Bottin, F.; Bouchet, J.; Bousquet, E. The ABINIT Project: Impact, Environment and Recent Developments. *Comput. Phys. Commun.* **2020**, *248*, 107042. [CrossRef]
22. Romero, A.H.; Allan, D.C.; Amadon, B.; Antonius, G.; Applencourt, T.; Baguet, L.; Bieder, J.; Bottin, F.; Bouchet, J.; Bousquet, E. ABINIT: Overview and Focus on Selected Capabilities. *J. Chem. Phys.* **2020**, *152*, 124102. [CrossRef]
23. Bellaiche, L.; Vanderbilt, D. The Virtual Crystal Approximation Revisited: Application to Dielectric and Piezoelectric Properties of Perovskites. *Phys. Rev. B* **2000**, *61*, 7877–7882. [CrossRef]
24. Ghosez, P.; Desquesnes, D.; Gonze, X.; Rabe, K.M. First-Principles Study of Lattice Instabilities in $Ba_xSr_{1-x}TiO_3$. *AIP Conference Proceedings* **2000**, *535*, 102–110.
25. Perdew, J.P.; Burke, K.; Ernzerhof, M. Generalized Gradient Approximation Made Simple. *Phys. Rev. Lett.* **1996**, *77*, 3865–3868. [CrossRef]
26. Hashi, K.; Ishikawa, K.; Matsuda, T.; Aoki, K. Microstructure and Hydrogen Permeability in Nb–Ti–Co Multiphase Alloys. *J. Alloys Compd.* **2006**, *425*, 284–290. [CrossRef]
27. Yan, E.; Li, X.; Rettenmayr, M.; Liu, D.; Su, Y.; Guo, J.; Xu, D.; Fu, H. Design of Hydrogen Permeable Nb–Ni–Ti Alloys by Correlating the Microstructures, Solidification Paths and Hydrogen Permeability. *Int. J. Hydrog. Energy* **2014**, *39*, 3505–3516. [CrossRef]
28. Ishikawa, K.; Tokui, S.; Aoki, K. Microstructure and Hydrogen Permeation of Cold Rolled and Annealed Nb40Ti30Ni30 Alloy. *Intermetallics* **2009**, *17*, 109–114. [CrossRef]
29. Santucci, A.; Tosti, S.; Basile, A. 4—Alternatives to Palladium in Membranes for Hydrogen Separation: Nickel, Niobium and Vanadium Alloys, Ceramic Supports for Metal Alloys and Porous Glass Membranes. In *Handbook of Membrane Reactors*; Basile, A., Ed.; Woodhead Publishing: Sawston, Cambridge, UK, 2013; Volume 1, pp. 183–217. ISBN 978-0-85709-414-8.
30. Yukawa, H.; Nambu, T.; Matsumoto, Y.; Watanabe, N.; Zhang, G.; Morinaga, M. Alloy Design of Nb-Based Hydrogen Permeable Membrane with Strong Resistance to Hydrogen Embrittlement. *Mater. Trans.* **2008**, *49*, 2202–2207. [CrossRef]
31. Ramachandran, A.; Zhuang, H.; Lackner, K.S. Probing the Interactions between Interstitial Hydrogen Atoms in Niobium through Density Functional Theory Calculations. *Mater. Today Commun.* **2020**, *25*, 101415. [CrossRef]

membranes

MDPI

Article

Reverse Osmosis Membrane Zero Liquid Discharge for Agriculture Drainage Water Desalination: Technical, Economic, and Environmental Assessment

Marwa M. El Sayed *, Abdelghani M. G. Abulnour, Shadia R. Tewfik, Mohamed H. Sorour, Heba A. Hani and Hayam F. Shaalan

Engineering and Pilot Plant Department, National Research Centre, Egypt El-Buhouth Street, Dokki, Cairo P.O. Box 12622, Egypt
* Correspondence: dr.marwameid@gmail.com; Tel.: +20-1096170061; Fax: +20-2-33370931

Abstract: Agricultural drainage water (ADW) represents a potential source for fresh water after receiving appropriate treatments to satisfy the water quality requirements. Desalination of ADW with medium salinity and moderate contamination with organic and inorganic chemical pollutants could provide a techno-economically feasible approach for facing water scarcity in arid areas. The current work presents a conceptual zero liquid discharge ADW desalination system proposed to treat 300,000 m^3/d. The system is based on pretreatment to remove impurities harmful to desalination by staged reverse osmosis (RO) membrane. The brine from the last RO stage is treated via thermal vapor compression followed by evaporation in solar ponds to recover more fresh water and salts of economic value. The essential technical features of the proposed system components are formulated. The proposed system components and its technical and economic indicators are deduced using available software for water pretreatment, RO membrane, desalination, thermal desalination, and solar evaporation ponds. The system provides total distilled water recovery of about 98% viz. 294,000 m^3/d in addition to recovered salts of 245,000 t/y. The net cost of water production amounts to USD 0.46 /m^3. The environmental considerations of the system are addressed and advantages of applying zero liquid discharge system are elucidated.

Keywords: agricultural drainage water; RO membrane; desalination; evaporation; economics; ZLD

1. Introduction

Water shortage has emerged as a major issue as the world's population grows and natural water supplies get depleted. It has been forecasted that the global demand for fresh water will exceed the supply by 40% by 2030 [1,2]. It is expected that water scarcity will increase from about one third to nearly half of the global urban population in 2050 [3]. Desalination has been widely used in regions with limited water supplies to address the problem of water shortages [4]. Desalination is the extraction of freshwater from brackish or salt water. Since the oceans cover 70% of the earth's surface, salty water desalination has the potential to offer a plentiful freshwater supply. Desalination has proved to be a viable and reliable solution to the worldwide water shortage during the last few decades. As of 2020, the global installed desalination capacities for freshwater production are about 97.2 million m^3/day provided by 20,971 projects [5]. The current global trend shows that membrane-based desalination technology is finding new outlets for supplying water to meet the growing water demand in most of the water-scarce countries [6,7]. The climate changes due to global warming are constantly increasing the salinity level of both land and seawater, thus reducing the availability of existing fresh water for households, agriculture, and industry. Irrigation with desalinated seawater has led to a significant increase in salinity and boron in the soils, which could affect the yield of moderately tolerant crops [8]. This has made it urgent to invent an appropriate water treatment technology that not only removes macro, micro, and nano-pollutants, but also desalinates water to a significant extent. Continued research and development of new treatment technologies are essential to improve the availability and quality of water supplies for agricultural use [9,10]. Consequently, thermal and membrane-based desalination technologies are playing an important role in solving global water scarcity problems.

Desalinated water has several advantages including (a) reduction of the possibilities of soil salinity, which has adverse effect on soil properties, (b) increasing the cultivated land area and the number of crops, and (c) improving crop quality and productivity [11]. Brackish water reverse osmosis (RO) and seawater reverse osmosis (SWRO) are the two most common desalination methods used to produce water for agriculture (SWRO). When compared to thermal desalination methods, SWRO has emerged as the most sophisticated and leading technique because of higher product water quality, reduced energy needs, and hence cheaper water costs [12].

Therefore, the aim of this work is to develop a cost effective and integrated scheme for ADW desalination with zero liquid discharge. The system is based on pretreatment to remove impurities harmful to desalination by staged reverse osmosis (RO) membrane.

2. Methods

2.1. Identification of Typical Characteristics of ADW

Agricultural drainage water (ADW) is characterized by variation of its composition from one site to another due to different effluent sources discharged to a specific drain [13]. Large drains or more specifically final drain streams exhibit moderate variation in quality as compared to small drains receiving frequent shock loads. Examples of large drains in Egypt are Bahr El-Bakar, El-Omoum, and Bahr Hadous. In this study, a typical main drain water source is considered, based on relatively stable water composition, allowing to propose an appropriate technique for treatment. Also, main drains allow considering large capacity and provide flexibility in the treatment plant site selection based on economic and environmental considerations.

Furthermore, a typical ADW treatment composition is considered. Table 1 presents the typical composition of ADW, which is mainly characterized by the following features: medium salinity in the range of brackish water, high hardness content, high suspended solids, low organic contamination, and low Fe and Mn content.

Table 1. Raw water composition.

Item	Na	K	Mg	Ca	Cl	HCO_3	SO_4	NO_3
Conc. (mg/L)	596	30	140	110	1070	370	355	40
Item	**Fe**	**Mn**	**TDS**	**TP**	**pH**	**TSS**	**COD**	**BOD**
Conc. (mg/L)	2	1	2705	1	7.7	80	20	10

2.2. Rationale of Developed Proposed Treatment Scheme

The proposed treatment system is based on technically solid concepts to be applied on large scale and is environmentally accepted. The developed system comprises the following sections:

Pretreatment includes a chlorination section followed by flocculation, filtration, and sludge dewatering. This process sequence ensures the removal of microbiological contamination (e.g., algae), and performs as a softener to remove most of the Ca and Mg content, Fe and Mn, suspended solids, and organic matters. Separated sludge is dewatered and dried to be sold as a recovered by-product, while filtrate from dewatering is recycled to the process. A dual media filtration stage ensures removal of residual turbidity.

Desalination is then proposed to produce high quality water for different applications. Reverse osmosis is the most economic technique applied for that purpose. Desalination is performed by two-stage reverse osmosis (RO) to achieve maximum water recovery. The previous pretreatment allows high water recovery from RO due to eliminating most of the impurities affecting RO performance.

The concentrate from RO is further desalted to increase freshwater recovery via thermal desalination technique to produce fresh water and concentrated brine.

The concentrated brine is directed to solar evaporation ponds to recover salts.

Recovered salts from solar ponds as well as dried residues from water pretreatment present an additional economic value.

3. Results

3.1. Basic Engineering of Proposed System

Basic engineering essentially comprises material balance and technical features as well as specifications for a specified system capacity. The basis of design of the system components is presented in Table 2.

Table 2. Basis of design of the system components.

	1. Pre-Treatment	
Raw Water Feed Flowrate		300,000 m^3/d
Chlorine dose		8 mg/L
Sodium hydroxide dose		400 mg/L
Flocculant		10 mg/L
System		Low pressure-pumping station, In-line chlorination, chemicals (caustic soda and flocculant) addition, flocculation, and clarification, in one tank, followed by dual media filtration.
Power		1525 kW
	2. RO stages	
	First Stage RO	Second Stage RO
Feed flow rate	300,000 m^3/d	30,000 m^3/d
Recovery	90%	60%
	3. Thermal Concentration	
Feed flowrate		12,000 m^3/day
System		Thermo vapor compression (TVC)
Power		913 kW
Recovery		50%
	4. Solar Pond	
Feed flow rate		6000 m^3/d
Average climate conditions:		
Temperature		24 °C
Humidity		50%
Solar irradiation		412.2 W/m^2
Wind speed		3.5 m/s
Rainfall		65 mm/y
Elevation above sea		15

The specifications of the main streams characteristics and components specification have been identified with the aid of software available for chemical pretreatment using the following software:

- WatPro, version 4.0, Hydromantis, water treatment simulator for predicting water quality, Hamilton, Ontario, Canada
- RO using LewaPlus®, version 5.0, Calculation and Design Software, LANXESS, Deutschland GmbH Liquid Purification Technologies Kennedy platz 150569 Cologne, Germany,
- Thermal desalination using WT Cost II© software developed by the Bureau of Reclamation and Moch Associates, U.S. Department of the Interior, Bureau of Reclamation Technical Service Center, Denver, Colorado, according to the design features provided, and solar evaporation ponds using Excel.

3.2. Material Balance

The material balance of the main components of the integrated ZLD desalination system is illustrated in Figure 1. Tables 3 and 4 present the water composition of the main pretreatment streams and RO stages, respectively.

Table 3. Water Composition of the main pretreatment streams.

	Raw Water	Sedimentation Effluent	Filtration Effluent
Na (mg/L)	596	945	945
Ca (mg/L)	110	28	28
Mg (mg/L)	140	46	46
K (mg/L)	30	30	30
Cl (mg/L)	1070	1070	1070
SO_4 (mg/L)	355	355	355
CO_3 (mg/L)	-	147	147
HCO_3 (mg/L)	370	232	232
NO_3 (mg/L)	40	40	40
Fe (mg/L)	2	-	-
Mn (mg/L)	1	-	-
TOC (mg/L)	10	1	-
TDS (mg/L)	2705	2896	2892
pH	7.7	9.8	9.8
TOC (mg/L)	10	1	-
Turbidity (NTU)	100	5	1
Solid ppt		$CaCO_3$ 205.6 mg/L	
		$Mg(OH)_2$ 228.6 mg/L	

Table 4. Composition of two-stage RO streams *.

Stream Item	Feed Raw Water	Permeate (1)	Permeate (2)	Concentrate (1)	Concentrate (2)
Na (mg/L)	945	17.5	130	9291	23,080
Ca (mg/L)	28	0.2	1.4	278.4	694
Mg (mg/L)	46	0.3	224	457.4	1140
K (mg/L)	30	0.83	6.12	292.7	722.5
Cl (mg/L)	1070	25	1863	13,622	33,772
SO_4 (mg/L)	355	1.3	9.7	3538.3	8831
CO_3 (mg/L)	147	0	0	0	0.003

Table 4. *Cont.*

Stream Item	Feed Raw Water	Permeate (1)	Permeate (2)	Concentrate (1)	Concentrate (2)
HCO_3(mg/L)	232	0.04	0.28	20.3	50.3
NO_3 (mg/L)	40	5.4	37.3	350.3	819
CO_2 (mg/L)	0.05	0	273.6	273.6	273.6
TDS (mg/L)	2892	273.6	373.3	24,850	69,108
pH	9.8	2.32	3.18	4.9	5.26

* Before post-treatment of permeate.

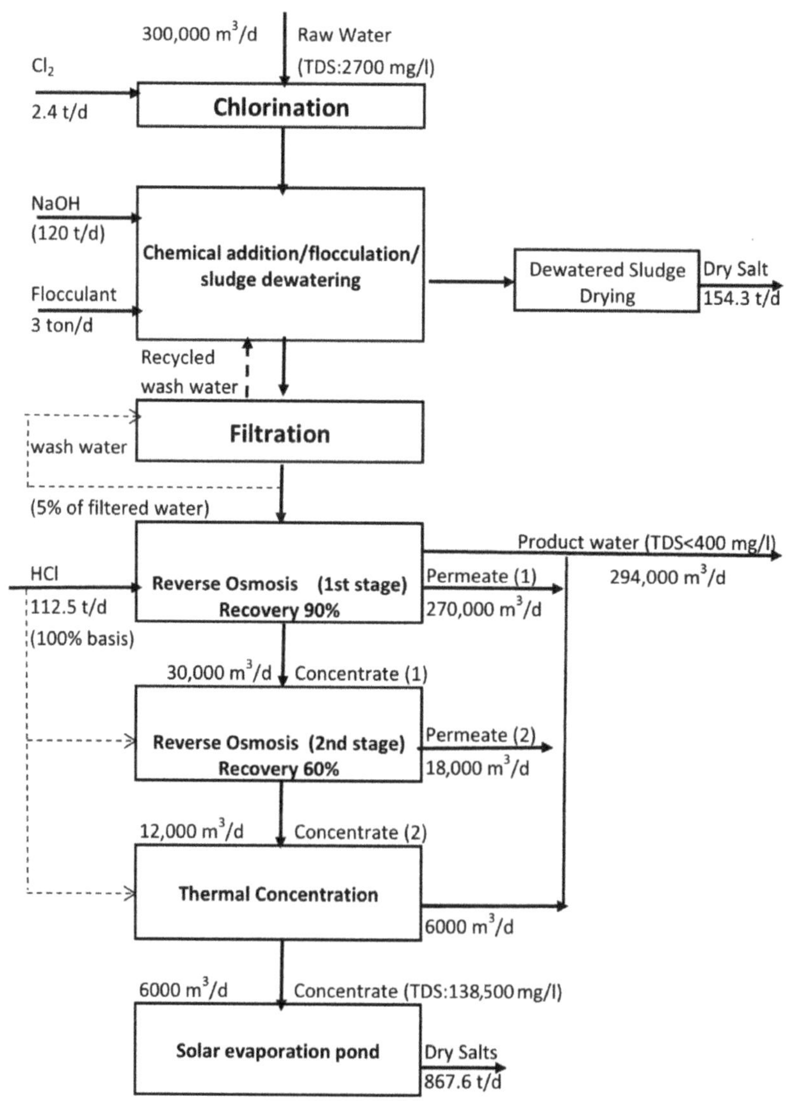

Figure 1. Process flow diagram, main components, and material balance of the proposed zero liquid discharge ZLD desalination system.

3.3. Technical Features and Specifications of the Proposed System

Table 5 presents the technical features and main technical specifications of the pretreatment stage, which comprises low pressure pumping station, chlorination, chemical treatment, flocculation and sedimentation, sludge dewatering, and clarified water dual media filtration.

Table 5. The main technical specifications of pretreatment components.

Item	Specifications
Chlorination	Chlorine gas vacuum feeding system 2 × 100 kg/h, with 10 chlorine gas cylinders (1 t capacity), with pumps and accessories, control and chlorine detection, and elimination system included in a separate building provided with proper ventilation.
Chemical handling system	NaOH storage tanks (3000) m^3 capacity, pumps centrifugal, CI, 20 m^3/h, 50 m head, feed tanks 100 m^3 capacity feed pumps CI 10 m^3/h, 20 m. Flocculant preparation tank (30 m^3) and pumps, (tanks 10 m^3 with mixers), feeding pumps 1 m^3/h, variable speed, 5 bar pressure.
Chemical treatment/flocculation/clarification tank	4 tanks each 40 m diameter, with 4 m side wall depth, with chemical addition/flocculation section inlet, discharge wires, and sludge discharge arrangements. Mixers for rapid mixing and flocculation and sludge scraper.
Sludge dewatering	The system includes sludge pumps 8 × 10 m^3/h, head 50 m. 4 sludge continuous belt filters, each 10 m^3/h. 4 belt conveyers for cake (1 m width, 20 m length). 4 filtrate recycling pumps 10 m^3/h, 20 m head.
Filtration unit	Dual media pressure filters (anthracite/sand). Operating pressure 3 bar. Diameter 3.5 m. Filtration rate 30 m^3/m^2 h. No. 50 With automatic backwash system.

The main technical features and specifications of two-stage RO are illustrated in Table 6. The first RO stage (RO1) is responsible for the recovery of fresh water up to 90% of feed water. The concentrate from this stage is directed to a second stage (RO2) to recover more fresh water of up to 60% of the quantity of the concentrate of RO1.

The concentrate of the second RO stage, (RO2) is treated in a thermal concentration unit based on thermo vapor compression (TVC) to recover up to 50% of the concentrate of RO2. The main system components are: boilers with accessories, distiller components, vessels and heat transfer tubes, pumps, degassing and chemical design units, piping including fittings and valves, electrical instrumentation and control, distiller's support including platforms and other necessary components, thermal ejectors, chemical cleaning system, and buildings.

The concentrate from thermal concentration is directed to solar evaporation ponds in a nearby environmentally, economically, and socially accepted site to evaporate water and recover salts of economic value. The effective area of ponds amounts to 401,630 m^2. The actual area includes banks and walkways, which amount to 50 m^2. The walls and bottom are lined with layers of clay, geotextile, and HDPE sheets to mitigate any salty water leakage to ground water.

Table 6. Technical specification of RO units.

Item	Specifications	
	1st Stage	**2nd Stage**
Feed flow rate	300,000 m^3/d	30,000 m^3/d
Recovery	90%	60%
Feed pressure "bar"	28.7	57.0
Pump pressure "bar"	5	5
Concentrate pressure "bar"	30	30
High pressure pump Pump/ motor efficiency	84%/94%	84%/94%
High pressure pumps motors power	12,622 kW	2502 kW
Pump/motor efficiency	84%/94%	84%/94%
Booster pump motors power.	626 kW	-
System components	Cartridge filters, RO trains, RO membrane elements, high pressure pumps with energy recovery. Inter-connecting piping, post treatment, membrane cleaning unit, electrical system, and instrumentation and control, auxiliary equipment and building.	
	System Configuration	
Number of stages/elements	2/13,150	1/1250
Elements number per vessel	SW HR 400 34/10	SW HR 400 34/10

4. Financial Indicators for ZLD System

4.1. Cost Estimation and Financial Indicators

Cost estimates have been performed based on updated published data to the year 2021 using Chemical Engineering Cost index and the prementioned software in Section 3.2

4.2. Construction Cost Estimates

The summary of the construction cost of the proposed system is presented in Table 7. The construction cost is estimated to be about USD 116.4 million. Reverse osmosis units' costs represent about 51.4% of the total costs. The total desalination cost (including RO and TVC) amounts to USD 74.92 million, representing about 64% of the total construction cost. The estimated construction costs per m^3/day amounts to USD 396/m^3d. These findings agree well with published work where the capital cost for brackish water RO desalination was USD 100 million at 300 m^3/d by adopting a developed empirical correlation from hundreds of desalination plants [14].

Table 7. Summary of construction cost estimates.

No	Item	Cost USD 1000
1.	Pumping station	11,394
2.	Pretreatment Chlorination Chemical treatment/flocculation clarification and sludge dewatering Filtration	 2160 6933 5398
	Total pretreatment	25,885

Table 7. *Cont.*

No	Item	Cost USD 1000
3.	Reverse osmosis units RO1 RO2	 56,210 3610
	Total RO units	59,820
4.	Thermal concentration	15,100
5.	Solar evaporation	15,600
	Total	116,405

4.3. Operating Costs Estimates

Table 7 presents a summary of the operating costs estimates for the proposed system. The annual operating cost is estimated to be about USD 48 million. The major cost items are chemicals and electricity, representing 53% and 21.5%, respectively.

4.4. Annual Production Costs Estimates

The annual production costs are the results of the annual operating cost in addition to annual capital amortization.

Tables 8 and 9 present the estimated operating costs, revenues, and total production costs.

Table 8. Estimates of annual operating costs.

No	Item		Annual Cost USD 1000/y
1.		Electricity	10,290
		Electricity consumption: 1.47×10^8 kWh/y @ USD 0.07/kW/h	
2.	Fuel	75 t/d @ USD 100/t	247
3.	Chemicals	NaOH 120 t/d @ USD 300/t Cl_2 2.4 t/d @ USD 150/t HCl (100% basis) 112.5 t/d @ USD 300 t/d Flocculant/anti-scaling 3.6 t/d @ USD 2000/t	11,826 118.3 11,087 2365.2
			25,396.5
4.	Membrane Replacement	Membrane lifetime 3 y Membrane cost USD 400/element	1920
5.	Maintenance	3% of capital cost	3492
6.	Labor	150 persons @ USD 15,000/y	2250
	Subtotal		43,595.5
7.	Other operating cost	10% of subtotal	4359.5
		Total operating cost	47,955.0

The net annual production costs are estimated to be about USD 56 million. Considering the revenues from recovered salts, the net annual production costs are estimated to be about USD 43.7 million. The net production cost per cubic meter of produced water is USD 0.453 m^3.

The published range for the unit cost for brackish water RO desalination for 10,000 to 70,000 m^3/d capacities and 2000–6000 mg/L are 0.39–0.66 USD/m^3 [15,16]. It is worth mentioning that using electrodialysis for brackish water desalination may be lower by 5–33% according to the capacity (Generous et al., 2021). In general, the RO desalination unit cost for seawater and brackish water for capacities up to 450,000 m^3/d is considered to be in the range of 0.7–1.4 USD/m^3 and 0.3–0.7 USD/m^3, respectively [14]. Moreover, Panagopoulos (2021) studied brackish water and seawater ZLD systems. He concluded that for the seawater system, the estimated cost is 0.84 USD/m^3 which is 1.24 times higher than the brackish [17].

Table 9. Estimated Annual Costs, Revenues and Cost/m^3..

Item	Basis	Annual Cost USD 1000
Operating cost	According to Table 8	47,955
Capital amortization	Plant lifetime 20 y Interest rate 4%	8566
Total annual cost		56,521
Revenues	Mixed salts (NaCl, $CaCl_2$, $MgSO_4$) 245,000 t/y @ USD 50/t	12,250
Net annual cost		44,271
Cost/m^3	Plant operation factor 0.9	USD 0.46

5. Environmental Considerations

The following environmental issues are addressed.

The zero-discharge desalination system provides many advantages related to environment. These include maximization of natural resources recovery and prevention of ground water contamination.

The site of the plant should be carefully selected on the bank of a main drain away from discharge points from polluted sources to avoid shock loads that sharply affect the quality of the water fed to the plant, which may cause severe deterioration of plant performance.

The site of the evaporation ponds should also be properly selected. It should be far from the residential and agricultural areas. It is preferably in a desert area near the plant site. The nature of the site and its hydrogeological characteristics should allow for environmental and economic ponds construction.

Sludge resulting from pretreatment containing mainly calcium and magnesium salts could be safely disposed of or reused after natural sun drying.

The costs of the proposed system should be re-evaluated based on the actual site conditions, which affect production cost, especially chemical water treatment, which represents a considerable proportion of the cost.

6. Conclusions

Agricultural drainage water (ADW) with medium salinity and heavy metal contamination could be treated to produce water of high quality for reuse. Zero liquid discharge desalination is technically and environmentally an accepted approach, by attaining maximum water and salt recovery for economic applications.

This case under study for treating ADW with a capacity of 300,000 m^3/d concluded a total water recovery of 98% of the raw ADW. The 2% concentrated brine treated in solar evaporation ponds could recover about 245,000 ton/per year of salts, containing mainly sodium, potassium, calcium, and magnesium salts.

The estimated construction cost of the plant amounts to about USD 116.4 million, and the cost of produced water amounts to about USD 0.46/m^3.

The cost of water production could considerably increase if lower recovery values are considered. The brine disposal, as an alternative, although may not be accepted from the environmental point of view in many situations, yet, if accepted, could be applied in specific situations with potential cost reduction.

Author Contributions: Conceptualization, M.H.S. and M.M.E.S.; methodology, A.M.G.A. and H.F.S.; software, H.A.H.; validation; M.M.E.S., S.R.T.; investigation, M.H.S. and M.M.E.S.; original draft preparation, A.M.G.A.; review and editing, S.R.T.; funding acquisition, M.M.E.S. All authors have read and agreed to the published version of the manuscript.

Funding: This research received no external funding.

Institutional Review Board Statement: Not applicable.

Informed Consent Statement: Not applicable.

Data Availability Statement: Data available in a publicly accessible repository that does not issue DOIs. Publicly available datasets were analyzed in this study. This data can be found (https://www.davuniversity.org/images/files/study-material/PLANT%20DESIGN%20AND%20ECONOMICS%20FOR%20CHEMICAL%20ENGINEERS.pdf, accessed on 13 May 2022).

Acknowledgments: This work is funded by Science and Technology Development Fund, STDF, Ministry of Scientific Research, Egypt; Project No 41563 entitled "Advanced treatment of pre-treated wastewater using novel Nano-filtration/electro assisted adsorption system", 2020.

Conflicts of Interest: The authors declare no conflict of interest.

References

1. Water Resources Group. *Valuing Water, Enabling Change*; Annual Report; Water Resources Group: Washington, DC, USA, 2020.
2. UNU-INWEH. *Global-Water-Crisis-The-Facts*; Technical Report; UNU-INWEH: Hamilton, ON, Canada, 2017.
3. He, C.; Liu, Z.; Wu, J.; Pan, X.; Fang, Z.; Li, J.; Bryan, B.A. Future global urban water scarcity and potential solutions. *Nat. Commun.* **2021**, *12*, 4667. [CrossRef] [PubMed]
4. Nakoa, K.; Rahaoui, K.; Date, A.; Akbarzadeh, A. Sustainable zero liquid discharge desalination (SZLDD). *Sol. Energy* **2016**, *135*, 337–347. [CrossRef]
5. Thi, H.; Pasztor, T.; Fozer, D.; Manenti, F.; Toth, A.J. Comparison of Desalination Technologies Using RenewableEnergy Sources with Life Cycle, PESTLE, and Multi-CriteriaDecision Analyses. *Water* **2021**, *13*, 3023. [CrossRef]
6. Eke, K.; Yusuf, A.; Giwa, A.; Sodiq, A. The global status of desalination: An assessment of current desalinationtechnologies, plants and capacity. *Desalination* **2020**, *495*, 114633. [CrossRef]
7. Amy, G.; Ghaffour, N.; Li, Z.; Francis, L.; Linares, R.V.; Missimer, T.; Lattemann, S. Membrane-based seawater desalination: Present and future prospects. *Desalination* **2017**, *401*, 16–21. [CrossRef]
8. Kumar, R.; Ahmed, M.; Bhadrachari, G.; Jibu, P.T. Desalination for agriculture: Water quality and plant chemistry, technologies and challenges. *Water Sci. Technol. Water Supply* **2018**, *18*, 1505–1517. [CrossRef]
9. Matin, A.; Laoui, T.; Falath, W.; Farooque, M. Fouling control in reverse osmosis for water desalination & reuse: Current practices & emerging environment-friendly technologies. *Sci. Total Environ.* **2021**, *765*, 142721. [PubMed]
10. Shi, Y.; Zhang, C.; Li, R.; Zhuo, S.; Jin, Y.; Shi, L.; Wang, P. Solar evaporator with controlled salt precipitation for zero liquid discharge desalination. *Environ. Sci. Technol.* **2018**, *52*, 11822–11830. [CrossRef] [PubMed]
11. Nofal, I. Desal brackish water for agriculture. In Proceedings of the Regional Water Knowledge Sharing Forum, Sharm El Sheikh, Egypt, 12–14 May 2015.
12. Elimelech, M.; Phillip, W.A. The Future of Seawater Desalination: Energy, Technology, and the Environment. *Science* **2011**, *333*, 712. [CrossRef] [PubMed]
13. Florea, A.F.; Lu, C.; Hansen, H.C.B. A zero-valent iron and zeolite filter for nitrate recycling from agricultural drainage water. *Chemosphere* **2022**, *287*, 131993. [CrossRef] [PubMed]
14. Bhojwani, S.; Topolski, K.; Mukherjee, R.; Sengupta, D.; El-Halwagi, M.M. Technology review and data analysis for cost assessment of water treatment systems. *Sci. Total Environ.* **2019**, *651*, 2749–2761. [CrossRef] [PubMed]
15. Generous, M.M.; Qasem, N.A.A.; Akbar, U.A.; Zubair, S.S. Techno-economic assessment of electrodialysis and reverse osmosis desalination plants. *Sep. Purif. Technol.* **2021**, *272*, 118875. [CrossRef]
16. Pearson, J.L.; Michael, P.R.; Ghaffour, N.; Missimer, T.M. Economics and Energy Consumption of Brackish Water Reverse Osmosis Desalination: Innovations and Impacts of Feedwater Quality. *Membranes* **2021**, *11*, 616. [CrossRef] [PubMed]
17. Panagopoulos, A. Energetic, economic and environmental assessment of zero liquid discharge (ZLD) brackish water and seawater desalination systems. *Energy Convers. Manag.* **2021**, *235*, 113957. [CrossRef]

membranes

MDPI

Article

Photocatalytic Performance Improvement by Doping Ag on ZnO/MWCNTs Nanocomposite Prepared with Pulsed Laser Ablation Method Based Photocatalysts Degrading Rhodamine B Organic Pollutant Dye

Tahani A. Alrebdi [1], Reham A. Rezk [2], Shoug M. Alghamdi [3], Hoda A. Ahmed [4,5], Fatemah H. Alkallas [1], Rami Adel Pashameah [6], Ayman M. Mostafa [7,8,*] and Eman A. Mwafy [8,9]

1 Department of Physics, College of Science, Princess Nourah bint Abdulrahman University, Riyadh 11671, Saudi Arabia
2 Higher Technological Institute, 10th of Ramadan City, 6th of October Branch, 3rd Zone, 7th Section, 6th of October City, 10th of Ramadan 44629, Egypt
3 Department of Physics, Faculty of Science, Taibah University, Yanbu 46423, Saudi Arabia
4 Department of Chemistry, Faculty of Science, Cairo University, Cairo 12613, Egypt
5 Chemistry Department, College of Sciences, Taibah University, Yanbu 46423, Saudi Arabia
6 Department of Chemistry, Faculty of Applied Science, Umm Al-Qura University, Makkah 24230, Saudi Arabia
7 Spectroscopy Department, Physics Division, National Research Centre, 33 El Bohouth st. (Former El Tahrir st.), Dokki, Giza 12622, Egypt
8 Laser Technology Unit, Center of Excellent for Advanced Science, National Research Centre, 33 El Bohouth st. (Former El Tahrir st.), Dokki, Giza 12622, Egypt
9 Physical Chemistry Department, Advanced Materials Technology and Mineral Resources Research Institute, National Research Centre, Dokki, Giza 12622, Egypt
* Correspondence: aymanmdarwish@gmail.com

Abstract: ZnO/MWCNTs nanocomposite has significant potential in photocatalytic and environmental treatment. Unfortunately, its photocatalytic efficacy is not high enough due to its poor light absorbance and quick recombination of photo-generated carriers, which might be improved by incorporation with noble metal nanoparticles. Herein, Ag-doped ZnO/MWCNTs nanocomposite was prepared using a pulsed laser ablation approach in the liquid media and examined as a degradable catalyst for Rhodamine B. (RhB). Different techniques were used to confirm the formation of the nanostructured materials (ZnO and Ag) and the complete interaction between them and MWCNTs. X-ray diffraction pattern revealed the hexagonal wurtzite crystal structure of ZnO and Ag. Additionally, UV-visible absorption spectrum was used to study the change throughout the shift in the transition energies, which affected the photocatalytic degradation. Furthermore, the morphological investigation by a scanning electron microscope showed the successful embedding and decoration of ZnO and Ag on the outer surface of CNTs. Moreover, the oxidation state of the formed final nanocomposite was investigated via an X-ray photoelectron spectrometer. After that, the photocatalytic degradations of RhB were tested using the prepared catalysts. The results showed that utilizing Ag significantly impacted the photo degradation of RhB by lowering the charge carrier recombination, leading to 95% photocatalytic degradation after 12 min. The enhanced photocatalytic performance of the produced nanocomposite was attributed to the role of the Ag dopant in generating more active oxygen species. Moreover, the impacts of the catalyst amount, pH level, and contact time were discussed.

Keywords: nanocomposite; NPs; water treatment; dye; laser ablation; Rhodamine B

1. Introduction

Due to the industry's regular activities, dye pollution is one of the world's most serious environmental problems. Paper printing, textile dyeing, cosmetics, and pharmaceuticals are just few industries that use synthetic colors. About 20% of the entire world's dye manufacturing is wasted during the dyeing process. One of these dyes, Rhodamine B (RhB), a common dye used in industrial applications, can irritate the skin and eyes, as well as the gastrointestinal and respiratory tracts. Dye pollution is the easiest to discover because it can be noticed by the naked eye at very low levels. Furthermore, color inhibits the passage of light into the water, which has an adverse effect on the photosynthetic process that supports the living organisms' environment under the water [1–5].

To overcome these hazard compounds, physical adsorption, chemical oxidation, and reverse osmosis membrane treatment processes are some of the physical and chemical techniques that can successfully remove dye pollutants from water. However, the cost of these techniques is very high, and there are some practical issues, such as the need to use many chemicals and incompletely degraded products produced during the oxidation process; therefore, these methods cannot be used to treat actual dye wastewater on a wide scale due to numerous practical issues, such as the disposal of a lot of adsorbent residue, expensive membrane treatment equipment, and operation costs. The catalytic reduction process shows a great achievement in the field of water treatment with the assistance of nanostructured materials for a fast reduction or oxidation reactions. This method is based on converting the hazard's organic form to a nontoxic form that is capable of being safely degradable in the water without causing any side effects to the environment. When nanostructured materials are used as catalytic degradable materials, two ways could be used to achieve the reduction. One of these ways is the photocatalytic degradation method, which is based on generating powerful hydroxyl (OH) and superoxide ($^{\bullet}O_2$) radicals. These radicals can non-selectively oxidize all types of organic molecules present in wastewater. The other method is the oxidation reduction catalytic degradation method. The photocatalytic technique has emerged as one of the possible processes for treating dye wastewater due to the benefits of low cost and ease of use. Therefore, it will be crucial to create new catalysts able to absorb sunshine or other forms of visible light that are inexpensive, simple to prepare, and available in a variety of materials in order to address the energy crisis and advance pollution prevention [6–12].

Recently, there have been significant advances in the field of nanotechnology, many of which have speeded it up. According to their types, procedures, morphology, size, crystal structures, and other characteristics, researchers are generating a diversity of nanomaterials with improved properties. Metals, their oxides (MOs), nano-graphenes, carbon nanotubes, and nano-composites are the most important nanomaterials that may be manufactured in a variety of ways with extreme precision. These nanomaterials have incredibly controlled size, larger surface area, easy fabrication methods, and unique features and uses, and they could be prepared chemically, physically, and through biological approaches [13–17]. For the degradation of organic dyes, several metal oxide semiconductors have been utilized, including titanium dioxide, ferric oxide, cadmium oxide, magnesium oxide, zinc oxide (ZnO), etc. Since ZnO has a wide band gap (3.37 eV) and a high excitonic binding energy, it has been regarded as an effective and non-toxic photocatalyst. Due to its safe performance, nanostructured ZnO has been widely employed in bio-imaging, drug delivery, antibacterial, biosensors, chemosensors, anticancer, solar cell applications, and photocatalysts. ZnO is a potential material for photocatalytic degradation of water contaminants in the field of water treatment, having exceptional activity, low cost, and environmentally acceptable features. After being exposed to photons with energy larger than or equal to its band gap, nanostructured ZnO suggests that reactive oxygen species are formed as a result of the interaction of charge carriers (excitons) with dissolved oxygen and water molecules. However, due to the decrease in the generation of species brought on by the simpler recombination of excitons, the photocatalytic activity of un-doped ZnO-NPs remains incredibly low [18–21].

Several crucial attempts have been made to improve the photocatalytic performance of ZnO-NPs by lowering the exciton recombination and, as a result, the creation of super-reactive oxygen species. Several strategies have been developed to improve nanostructured

ZnO's photocatalytic performance by producing more of these species. Because it traps electrons and intensifies the separation of charge carriers, doping metallic ions with ZnO lattice is one of the intriguing methods for enhancing photocatalytic performance. Noble metals, such as silver (Ag), have been favored for the doping of ZnO because of their high reduction potential and improved photo-degradation capabilities. Because of its high reduction potential, doping of ZnO with Ag is best suited for preventing the recombination of charge carriers by trapping photo-excited electrons from the conduction band (CB), increasing the number of species that react with the pollutant and destroy it [14,22–24]. Furthermore, the nanocomposite faced another problem that came from the hindered aggregation effect that was produced by the high amounts of the nanostructured materials, leading to a decrease in the dangling bonds and reactivity, consequently reducing the photocatalytic performance. Therefore, the use of matrix material with a surface-to-volume ratio, such as carbon nanotubes (CNTs), could represent the optimum solution for solving this problem. [25–30].

Different techniques could be used to synthesize nanostructured materials to create nanocomposite structures. Pulsed laser ablation in liquid media (PLAL) is an effective, simple, and environmentally friendly process that can create NPs with a wide range of sizes and functionalities by only changing the laser's parameters and the type of liquid medium. PLAL has been used to create a wide variety of nanoparticles, proving their adaptability. The benefits of this process include its simplicity, low cost, lack of vacuum champers, and ability to produce NPs without contamination in an incredibly clean and dependable manner. To produce a tiny particle with a distinct feature at the nanoscale, the PLA approach relies on the employment of high-intensity short pulses of up to a nanosecond from an infrared laser source in a concentrated environment [31–38].

In this article, we used the *eco*-friendly pulsed laser ablation in liquid media technique to synthesize ZnO nanoparticles and decorate CNTs to create the ZnO/MWCNTs nanocomposite, followed by using the same method to generate Ag nanoparticles interacting with ZnO/MWCNTs forming Ag-doped ZnO/MWCNTs. After that, XRD, SEM, EDX, XPS, and UV-Vis studies were used to characterize the produced NPs and nanocomposites. The ability of the created nanocomposite structure to eliminate organic pollutants, such as Rhodamine B, was next examined. For that, the novelty of the work was based on the use of a simple, *eco*-friendly, and green method to generate two different types of nanostructured materials (semiconducting ZnO and noble metal Ag) and embedding them to the matrix of CNTs to form the Ag-doped ZnO/MWCNTs nanocomposite. This combination between semiconductors and noble metals allows for enhancing the photocatalytic activity against organic pollutants, such as Rhodamine B.

2. Materials and Experimental Work

2.1. Preparation of Ag-Doped ZnO/MWCNTs Nanocomposite

Zn pellet (Zn, BDH chemical Ltd. pool in UK) was used for the synthesis of ZnO nanostructured by pulsed laser ablation of cleaned Zn metal targets placed in a glass vessel filled with 5 mL of a solution of functionalized MWCNTs (f-MWCNTs) to produce the ZnO/MWCNTs nanocomposite, followed by repeating the previous procedure of pulsed laser ablation of another target, silver pellet (Ag, BDH chemical Ltd. pool in UK), immersed in the produced solution of the ZnO/MWCNTs nanocomposite. The functionalization procedure of MWCNTs was previously mentioned in the work of A. Mostafa et al. [39,40]. During this process, the laser beam of the Nd: YAG laser (λ = 1064 nm) with a 7 ns, a 10 Hz repetition rate, and 150 mJ of laser energy was focused on the surface of the target via convex lens for 10 min, as shown in the schematic diagram of the casting work in Figure 1.

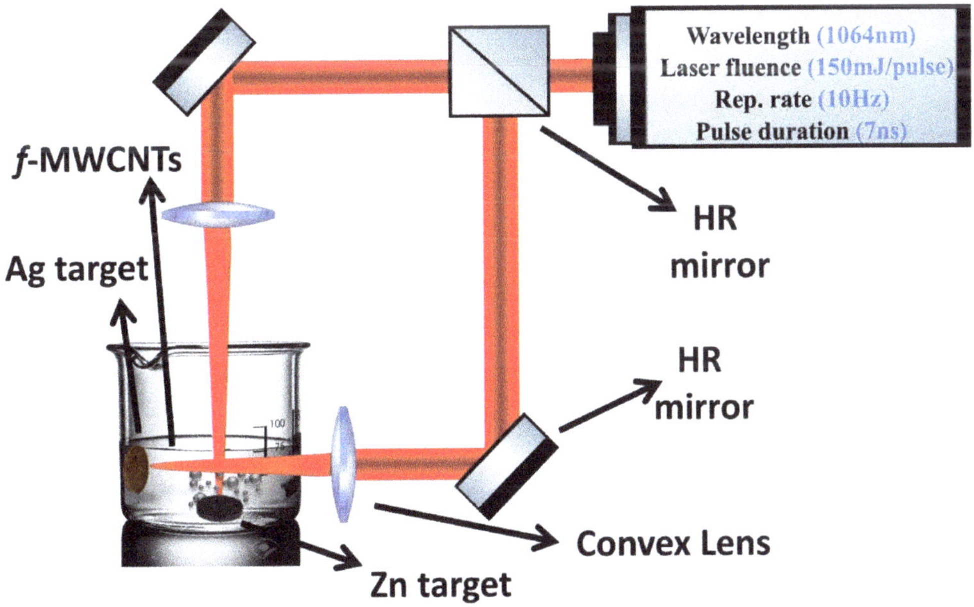

Figure 1. Schematic structure of preparation of Ag-doped ZnO/MWCNTs nanocomposite.

2.2. Determination of Concentration of Nanostructured Materials

The amount of generation of the nanostructured material of ZnO or Ag can be simply estimated by using the following procedure. PLAL method was used to generate Ag NPs through ablation of an Ag plate. The amount of Ag NPs was related to the weight loss in the primary Ag's precursor. So, the total number of Ag NPs was calculated as follows:

$$\text{Weight of generated Ag} = \text{Weight of Ag}_{(\text{before PLAL})} - \text{Weight of Ag}_{(\text{after PLAL})}$$

By knowing the volume of the solution (10 mL), the concentration of generated Ag NPs can be simply calculated as 8.2 μg/L.

The generated nanocomposite was collected, cleaned with ultra-pure water, filtered, and dried at 50 °C to guarantee only calculated silver was linked to CNTs. The nanocomposite was immersed in 15 mL of 70% nitric acid to quantify the total amount of coated Ag NPs on CNTs. So, Ag^+ concentration could be calculated. Based on Ag NPs only on the CNTs' outer surface, the loading amount of Ag in CNTs (W_{Ag}) was determined by:

$$\text{Weight}_{\text{decorated Ag on CNTs}} = \frac{C_{Ag+}V}{M - C_{Ag+}V}$$

where C_{Ag+}, V, and M are the concentration of Ag^+ in the solution, the volume of the solution (250 mL), and the weight of the produced nanocomposite (5 mg) [41]. The concentration of Ag nanoparticles decorated on CNTs showed the decorated Ag NPs on the nanocomposite could be simply calculated as 7.7 μg/L.

2.3. Characterization Techniques

The crystalline structure of the prepared samples was characterized using an X-ray diffractometer (Shimadzu XRD 7000, Kyoto, Japan). The absorbance spectra of the prepared materials were studied by a UV–VIS-NIR spectrophotometer (JASCO 570 UV–VIS-NIR, Tokyo, Japan). The identification of the oxidation state was carried out by an X-ray photoelectron spectrometer (Thermo Scientific K-ALPHA instrument, East Grinstead, UK). The morphological analysis was carried out by a scanning electron microscope with a beam voltage of 20 kV associated with an EDX analyzer detector (PHILIPS/FEI QUANTA 250, Prague, Czech Republic) to study the chemical composition using TEAM® software.

2.4. Adsorption Study

The degradation product, Rhodamine B (RhB, BDH Chemicals Ltd. pool in England), was used as a model pollutant for the degradation process, and the product's photocatalytic activity was examined. As a UV light source in this investigation, a 40 W Xe lamp was used. In a typical experiment, the RhB solution was mixed with the photocatalyst. The suspension was discovered under light illumination after a 60 min dark adsorption period. The remaining RhB concentration was then measured using UV-visible absorption spectroscopy at 554 nm. Additionally, utilizing an appropriate amount of the photocatalyst (10–50 mg/L) with pH (pH = 3–11), contact time up to 140 min, and the starting concentration of RhB (5–110 mg/L), the photocatalytic performance of Ag-doped ZnO/MWCNTs was determined. Each experiment with a tested photocatalyst was repeated three times to evaluate the experimental errors. The following equation was used to compute the removal extent of RhB [42–44]:

$$\text{Removal extent, percentage} = (C_o - C_a)/C_o \times 100 \quad (1)$$

where C_o and C_a are the concentration of the adsorbent material at the beginning of the adsorption process and after reaching the equilibrium of the adsorption process, respectively.

2.5. Desorption Study

To evaluate the reusability of native and modified sorbents in a successive adsorption and desorption process for a good number of recycles, we first carried out the RhB sorption under batch settings and then performed the desorption tests using 1 M of sodium hydroxide (NaOH) as the eluting agent. As with the sorption process, the desorption experiment begins with the addition of an appropriate amount of oven-dried sorbents to 100 mL of 1 M NaOH desorbing solution and then continues under conditions that are very similar to those of the sorption process. After that, the eluting agent was washed out of the regenerated sorbent with distilled water, and the sorbent was dried in a hot air oven. Once again, the same steps were used for a sorption–desorption cycle that may be repeated up to five times.

3. Results and Discussion

3.1. Investigation of the Prepared Ag NPs Doped ZnO/MWCNTs Nanocomposite

To verify the crystallinity and crystallite size, the XRD patterns of MWCNTs, ZnO/MWCNTs, and Ag NPs doped ZnO/MWCNTs nanocomposites were measured. The graphite structure of the MWCNTs is shown in Figure 2 along with how it developed in the diffraction patterns following embedding with ZnO NPs and then Ag NPs. The patterns demonstrate that the graphite structure coincided with the interplanar (002) of CNTs when the diffraction peaks at 26.10° occurred based on XRD card No. (PDF # 41-1487) [45]. Additionally, the diffractogram showed that MWCNTs decorated with ZnO NPs exhibit sharp peaks at 32.12°, 34.63°, 36.46°, 47.61°, 56.67°, 62.89°, 66.11°, 67.93°, and 69.01°. These diffraction peaks' positions indicate the hexagonal wurtzite ZnO structure, since they correlate with the (1 0 0), (0 2 2), (1 0 1), (1 0 2), (1 1 0), (1 0 3), (2 0 0), (1 1 2), and (2 0 1) crystal planes, respectively, based on JCPDS card No. (36-1451) [46]. The diffractogram showed patterns with no additional diffraction peaks rather than a graphite structure and a ZnO structure. The sample was free of any impurities. The diffraction data clearly show that ZnO NPs had good crystallinity, with a predicted Scherer formula crystallite size of 22.5 nm. In other words, when Ag NPs were produced and doped with ZnO/MWCNTs, new diffraction peaks appeared at 38.41°, 44.47°, and 64.32°, respectively, corresponding to (111), (200), and (220), respectively, of the crystal faces of the silver structure, which was consistent with JCPDS card No. (65-2871) of the hexagonal wurtzite of the Ag structure [47].

It was concluded that the hybrids of Ag NPs doped ZnO/MWCNTs maintained the new crystal structure of the metal Ag, ZnO, and carbon nanotubes, which would be essential to guarantee the distinctive crystal structure and outstanding performance.

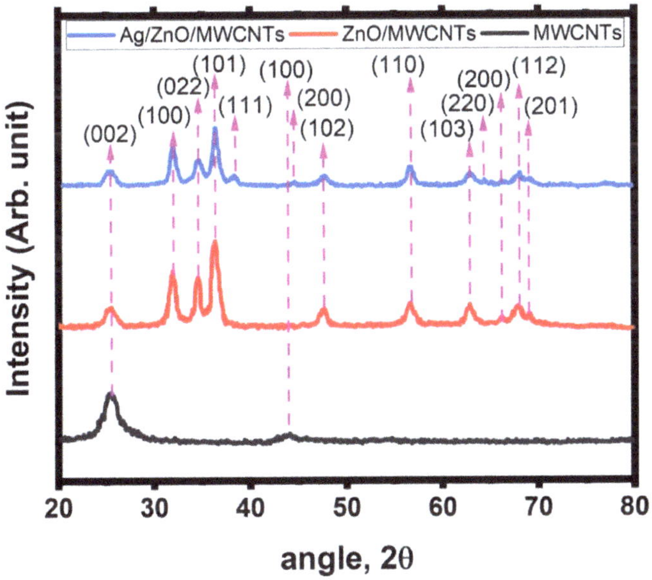

Figure 2. XRD diffractogram of MWCNTs, ZnO/MWCNTs, and Ag-doped ZnO/MWCNTs nanocomposite.

Using XPS, we were able to learn about the nature of the bonds in an Ag-doped ZnO/MWCNTs composite, as shown in the XPS survey spectra of the as-prepared sample in Figure 3a, where the Zn, Ag, O, and C core level peaks can be detected [48]. Figure 3b–e display the high-resolution XPS spectra of Ag 3d, Zn 2p, C 1s, and O 1s, respectively. In Figure 3b, the Ag 3d peak was deconvolved into two peaks by Gaussian fitting of Ag $3d_{5/2}$ and Ag $3d_{3/2}$ with binding energies of 368.2 and 373.3 eV, respectively, suggesting the existence of silver in the composite [48]. Additionally, as can be seen in Figure 3c, the Zn 2p spectrum was deconvolved into two peaks by Gaussian fitting (Zn $2p_{3/2}$ and Zn $2p_{1/2}$), which emerged at 1021.8 and 1048.8 eV, respectively, suggesting the existence of zinc in the composite [48]. Moreover, the O 1s peak was deconvolved into three peaks by Gaussian fitting of C=O, C-OH, Zn-O, and ZnO-C, as shown in Figure 3d, which are located at 529.9, 530.2, 531.7, and 532.9 eV, respectively. The presence of the CNTs' backbone structure, as indicated by the first two tangled peaks, and the interaction of ZnO or Ag with the structure of the CNTs, as indicated by the third and fourth peaks, are both confirmed by the spectra. These XPS measurements, together with the microscopic findings, point to ZnO development occurring on the CNTS walls (presumably via the creation of Zn–O–C bonds) [49], as has been reported for metal/CNTs or metal oxide/CNT composite materials. Additionally, oxygen is likely directly bonded to the CNTS structure through the formation of strong covalent bonds between carbon and oxygen atoms (no Zn–C bonding was detected), as evidenced by the presence of oxygen components in the high-resolution XPS C 1s spectrum and the presence of carbon components in the O 1s spectrum [50,51]. According to Figure 3e, the high resolution of the C 1s peak was deconvolved into six peaks by Gaussian fitting located at 284.8, 285.6, 286.4, 287.1, 289.7, and 291.5 eV, which are attributed to the C-C, C-OH, C=C, O=C–O, C-O-Zn, and C-Ag groups. The first four peaks were related to the functionalized graphite structure of CNTs, while the last two peaks showed that ZnO and Ag interacted with the CNT skeleton [27,52,53].

Using their respective high-resolution XPS spectra, the amounts of the three precursor elements (C, O, Zn, and Ag) in the final composite structure were calculated. The atomic percentages of carbon, oxygen, zinc, and silver were found to be 40.67%, 32.22%, 18.17%, and 8.94%, respectively.

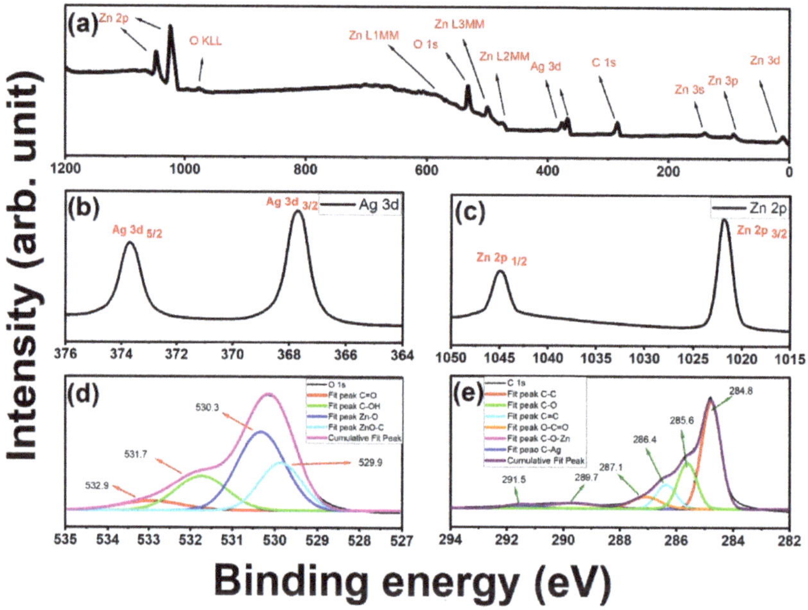

Figure 3. (**a**) A survey scan of XPS patterns of Ag-doped ZnO/MWCNTs nanocomposite and its HR spectra of (**b**) Ag 3d, (**c**) Zn 2p, (**d**) O 1s, (**e**) C 1s.

Figure 4 is the image of the SEM morphological analysis and elemental analysis of MWCNTs, ZnO/MWCNTs nanocomposite, and Ag-doped ZnO/MWCNTs. It shows that the image of the MWCNTs sample has smooth tubular shapes of the carbon nanotube structure, while the image of the ZnO/MWCNTs nanocomposite shows the tubular shape of MWCNTs decorated with highly distributed particles grown on their outer shape, whereas the image of the Ag-doped ZnO/MWCNTs nanocomposite shows that the amount of the distribution of particles on the tubular structure was increased, and the rate of roughness was increased, which was certainly related to the ablation of Ag metals and the embedding of another nanostructured material from the Ag nanostructured material. Moreover, the energy-dispersive X-ray (EDX) spectroscopy was used to examine the chemical purity and elemental composition of MWCNTs, ZnO/MWCNTs, and Ag doped with ZnO/MWCNTs. The EDX spectrum of the produced Ag NPs doped with ZnO/MWCNTs. The binding energies of Zn are associated, while the binding energies of Ag and oxygen are related to ZnO. Moreover, the presence of C is related to CNTs. Therefore, the EDX spectrum confirms that the synthesized hybrid structure contains the components Ag, Zn, C, and O [54,55].

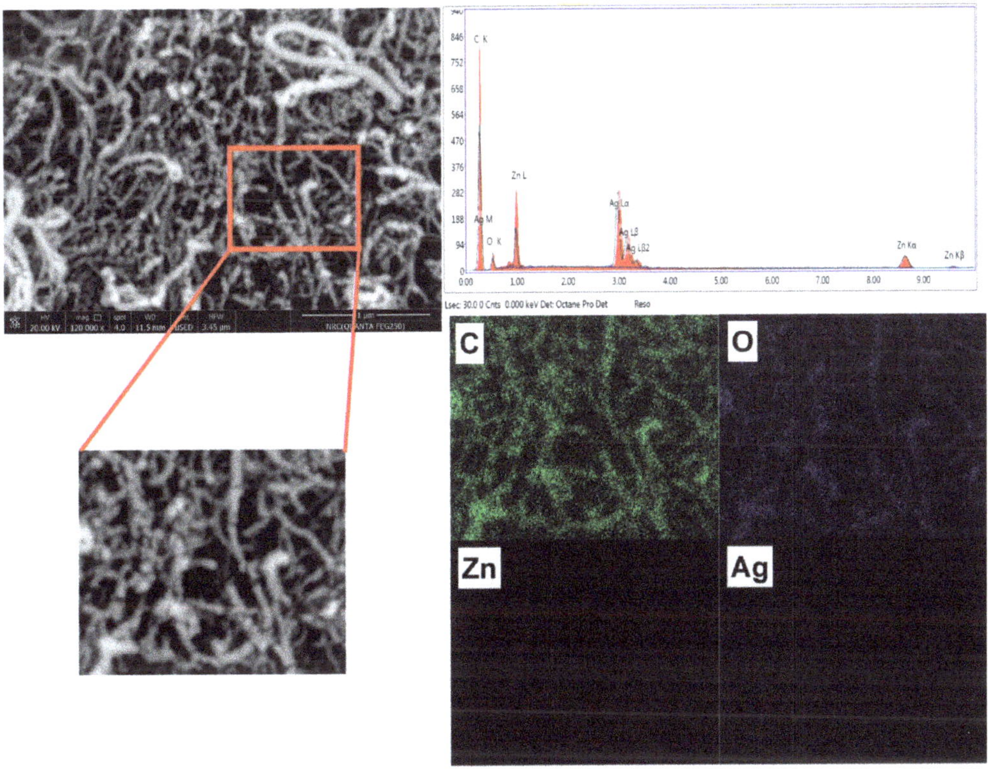

Figure 4. SEM image. EDX analysis and mapping of Ag-doped ZnO/MWCNTs nanocomposite.

The generated samples' optical characteristics have a significant impact on the photocatalytic activity. In this investigation, the effect of Ag NP doping and ZnO NP surface decoration on the optical characteristics of ZnO/MWCNTs was assessed using UV-VIS-NIR spectroscopy. The absorption spectra of three samples are shown in Figure 5a, showing Ag NPs doped with ZnO/MWCNTs exhibit the maximum light absorption in the 400–700 nm regions. The following equation can be used to calculate the products' band gaps [56]:

$$\alpha h\upsilon = A(\alpha h\upsilon - Eg)^n$$

where α, hv, A, and E_g are the absorption coefficient, photon energy, constant, band gap energy, respectively. For a hybrid structure to function as an indirect transition semiconductor, n must equal 1/2. To calculate the energy gap, the plot of $(hv)^{1/2}$ versus photon energy will be drawn. Pure ZnO/MWCNTs and ZnO/MWCNTs doped with Ag NPs had an estimated band gap of around 3.15 eV and 3.05 eV, respectively, as mentioned in Figure 5b. These findings support the doping of Ag NPs into the ZnO/MWCNTs matrix, since they demonstrate that adding Ag NPs causes the band gap of ZnO/MWCNTs to decrease. Based on the previous studies, this is because of the impurity that was introduced into the ZnO grains and may have trapped electrons stimulated from the conduction band, leading to a narrowing of the band gap and a continuous energy level [57].

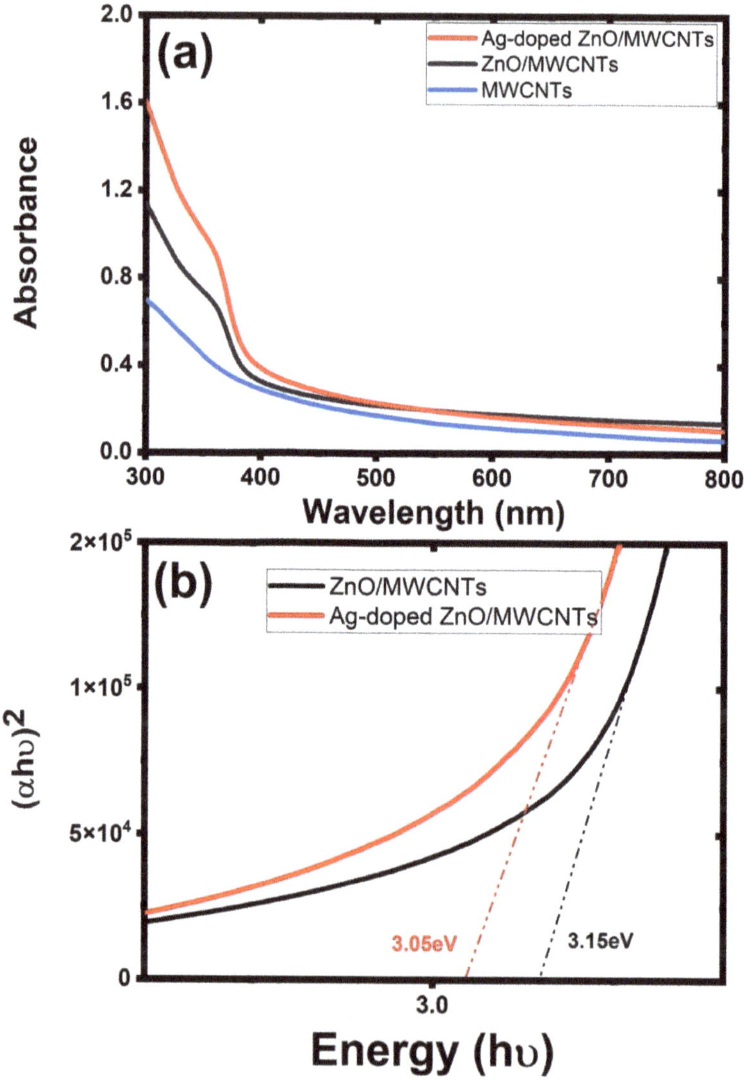

Figure 5. (**a**) Absorption spectrum of MWCNTs, ZnO/MWCNTs, and Ag-doped ZnO/MWCNTs nanocomposite and (**b**) their energy transition via Tauc relation.

3.2. Adsorption Process

3.2.1. Affected Parameters of the Adsorption Process

The following test circumstances and parameters were used for optimizing the photocatalytic process, as 10–50 mg of the catalyst, 3–11 pH of the solution, 0–140 min of the reaction contact time, 5–110 mg/L of the RhB concentration, and 100 mL of the RhB comprised the reaction conditions.

The outcomes are displayed in Figure 6a. RhB was rapidly removed when the catalyst concentration was increased from 10 mg to 50 mg, rising from 12.44% and 35.29% of 10 mg to 40.01% and 95.53% of 40 mg for ZnO/MWCNTs and Ag-doped ZnO/MWCNTs, respectively. As the elimination percentage rises, the amount of pollutants that can be degraded per unit mass of the catalyst progressively reaches its maximum, and as the catalyst concentrations rise further, the ability to improve degradation efficiency is severely constrained. Therefore, in this experiment, a catalyst dosage setting of 40 mg is ideal [11,12].

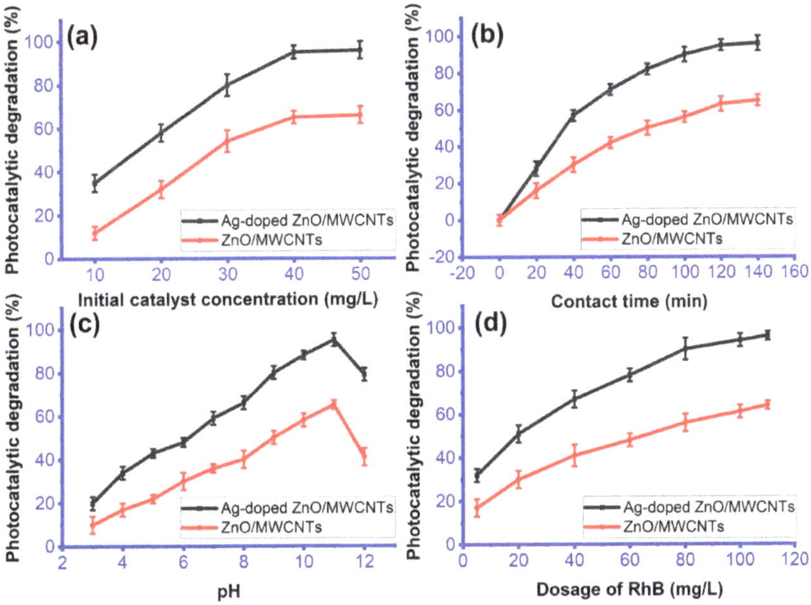

Figure 6. Effect of (**a**) initial catalyst concentration, (**b**) contact time, (**c**) pH, and (**d**) dosage amount of RhB on the removal of RhB.

In Figure 6b, by measuring the absorbance at 553 nm, the photocatalytic degradation of RhB utilizing the produced nanocomposite was assessed at various time intervals for 40 mg of the catalyst and at pH 10. With increasing irradiation time, a considerable decline in the RhB absorbance intensity at maximum 553 nm was seen, indicating a drop in dye concentration. According to a plot between the time and percent degradation efficiency of RhB, the deterioration effectiveness of RhB grows from 65.97% to 96.01% when the illumination time reaches 120 min and when the ZnO/MWCNTs decorated with Ag NPs is used [58–61].

According to research, industrial wastewater contains a wide range of pH values and plays a crucial role in the organic matter's ability to decompose. Therefore, it is crucial to assess the effectiveness of the catalyst at various pH levels. A catalyst's surface charge and the surface charge of organic dyes are both affected by pH, which also impacts how quickly the dye binds to the catalyst, and therefore, the rate of catalyst degradation. Therefore, using HCl and NaOH solutions to change the pH of the RhB solution, the impact of pH on the degradation process was investigated. Figure 6c demonstrates that the rate of degradation rose as the pH value increased, reaching a high (95%) at pH 11 (basic), starting to decline at pH 12 (80%), and reaching a minimum (20%) at pH 2 (acidic), which is also compatible with the literature. The catalyst and dye may repel each other electrostatically in acidic conditions where RhB occurs in cationic form (RhB^+), which reduces the effect of dye degradation due to RhB^+ being deprotonated when forming the zwitterion in basic media. So, the adsorption on the catalytic surface becomes difficult to be induced. In other words, at the fundamental conditions (above pH 11), the concentration of OH ions rises to the point where they cover the whole surface of the catalyst and leave it negatively charged. In this case, the adsorption process becomes easier due to the electrostatic attraction force produced from RhB^+ and the negatively charged surface of the catalyst. So, the attraction of the RhB by the catalyst surface increases. Moreover, the radical $^{\bullet}OH$ is capable of being easily generated by hydroxide ions on the surface of the nanocomposite, leading to a logical enhancement of the photocatalytic process [62–64].

Different initial dye concentrations were used to study the impact of the initial dye concentration on the degradation efficiency of the synthesized nanocomposite (Figure 6d). The findings demonstrated that the degrading efficiency of the catalysts increased with increasing initial RhB concentration from 33.21% to 95.12% for Ag-doped ZnO/MWCNTs, with an increase in RhB from 5 mg/L to 110 mg/L, whereas it was noted that the rate was slightly increased with the concentration increase, and this may be related to two factors: (i) as the dye concentration is increased, a high number of RhB molecules will accumulate on the catalyst's surface, preventing more of the dye molecules from reaching the surface; Second, (ii), when the dye molecules replicate, there will not be any available active sites for reaction. So, the increase rate would be slowed down [4,65–67].

3.2.2. Kinetic Study

The photocatalytic activity of ZnO/MWCNTs and their doping with Ag NPs to decompose RhB when exposed to light is depicted in Figure 7. After 120 min of ultraviolet exposure, compared to the pure ZnO/MWCNTs, Ag NPs doped MWCNTs have a higher photocatalytic degradation rate. MWCNTs, on the other hand, have a low photocatalytic activity in both regions. As a result, due to pure ZnO/MWCNTs having around 3.2 eV band gap, pure ZnO/MWCNTs can only be stimulated under UV light (376 nm). On the other hand, Ag NPs doped ZnO/MWCNTs exhibit improved photocatalytic capabilities in visible light, which is attributed to the narrowing of the band gap in comparison to pure ZnO/MWCNTs. Additionally, the photocatalytic activities of pure ZnO/MWCNTs and those doped with Ag NPs were compared using the results of the kinetic investigation. The reaction rate constants (k) can be determined by fitting the experimental data with the following pseudo-first-order reaction model [68–70]:

$$\text{Ln}\ (C_t/C_o) = kt$$

where the initial RhB concentration, the RhB concentration at time t, and the first-order rate constant are represented by C_o, C_t, and k, respectively. Figure 7 shows that the estimated data may be successfully fitted with the pseudo-first-order model. The rate constants for pure ZnO/MWCNTs and their decoration are 0.01098 min^{-1} and 0.03056 min^{-1}, respectively, indicating the more significant photocatalytic activity of Ag-doped MWCNTs in comparison to the pure MWCNTs, while the degradation of RhB dye is almost negligible under UV light irradiation in the absence of the photocatalyst, demonstrating the potential of ZnO/MWCNTs before and after embedding with Ag as a catalyst for the photocatalytic degradation of organic dyes. Moreover, the concentrations of RhB remain almost constant and almost negligible throughout the experiment in the presence of the photocatalyst (Ag-doped ZnO/MWCNTs) and in the absence of the catalyst [71]. The performance of the Ag-doped ZnO/MWCNTs nanocomposite in comparison with some representative materials reported in previous works was tabulated in Table 1.

Table 1. The catalytic degradation of different organic pollutants over other nanocomposite-based Ag/ZO/MWCNTs catalysts.

Pollutant	Catalyst	K (min^{-1})	Physical Shape	Ref.
phenol	1%CNT loaded with ZnO-Ag	0.0045	powder	[72]
phenol	5%CNT loaded with ZnO-Ag	0.0057	powder	[72]
phenol	10%CNT loaded with ZnO-Ag	0.0068	powder	[72]
phenol	20%CNT loaded with ZnO-Ag	0.0051	powder	[72]
Methylene blue (MB)	CNTs loaded ZnO/Ag	0.0282	powder	[25]
RhB	Ag-doped ZnO/MWCNTs	0.03056	powder	This study

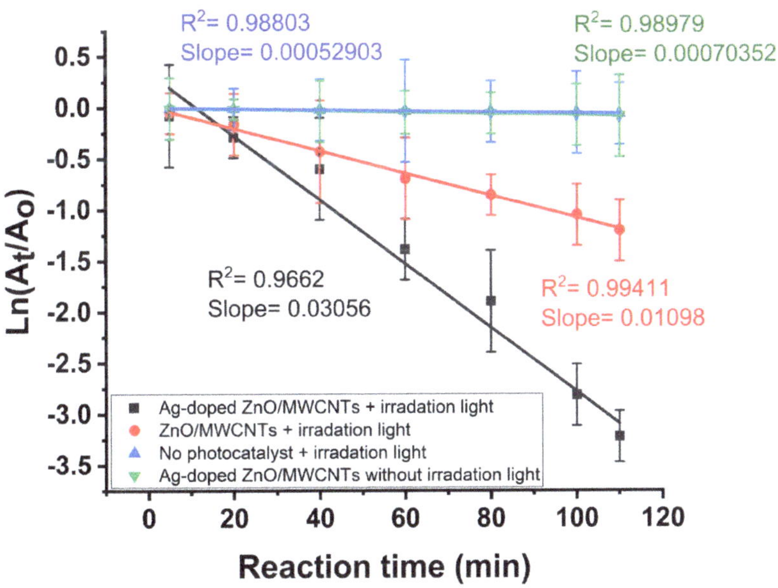

Figure 7. Absorption kinetic via pseudo-first order against RhB in the presence and absence of photocatalyst (ZnO/MWCNTs and Ag-doped ZnO/MWCNTs) and irradiation light.

3.2.3. Mechanism

Based on the experimental results, a suggested and schematic description of the photocatalytic activity of Ag NPs doped with ZnO/MWCNTs to destroy RhB under UV light was put forward. The pure ZnO/MWCNTs were unable to produce electron–hole pairs. The absorption of UV light generates electron–hole pairs in the nanocomposite particles. The holes react with the adsorbed H_2O molecules to create $^{\bullet}OH$, while the electron reduces O_2 and generates $O_2^{\bullet}$, H_2O_2, and $^{\bullet}OH$. The superoxide radicals ($^{\bullet}O_2$) and the photo-generated holes in the valence band (VB) have the potential to destroy RhB directly or indirectly by reacting with water to produce hydroxyl radicals ($^{\bullet}OH$). In addition, the photo-generated holes in the valence band (VB) have the potential to destroy RhB directly or indirectly by reacting with water to produce hydroxyl radicals ($^{\bullet}OH$) [73,74]. The radical-trapping experiment was mentioned in the schematic diagram of degradation of RhB in Figure 8.

3.2.4. Reusability

In addition, the photocatalyst's stability and usability are critical in real-world applications. As a result, after five consecutive cycles of UV irradiation (Figure 9), the photocatalytic performance of ZnO/MWCNTs with decorating Ag NPs was examined. The results showed that the photocatalytic performance of Ag NPs doped ZnO/MWCNTs was greater than 80% after five subsequent cycles, indicating its stability and reusability for practical applications. The losses in the degradation efficiency of the prepared nanocomposite in the reusability experiments were due to a decrease in the active sites on the nanocomposite's surface from using NaOH, which peeled its active areas, leading to the tendency of the catalyst to agglomerate, mass loss, and some disintegration during the recovery process of the catalyst [75].

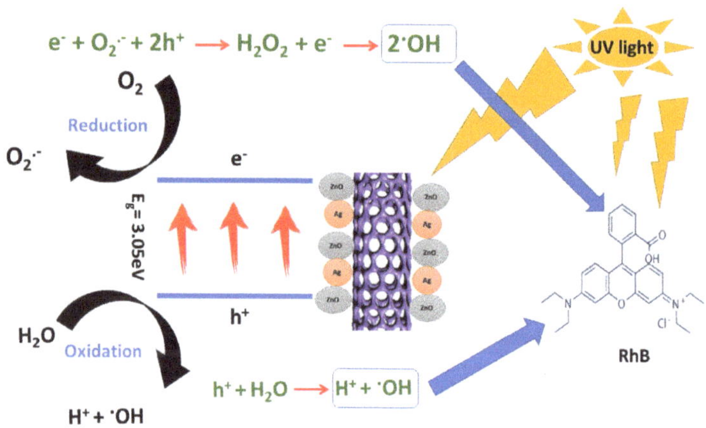

Figure 8. Schematic illustration of the photocatalytic mechanism of the photocatalyst for degradation of RhB.

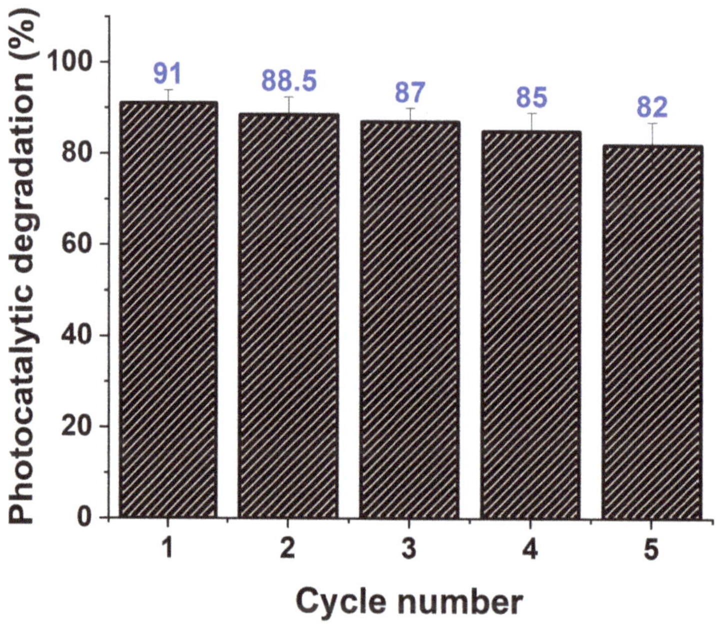

Figure 9. The usability of Ag-doped ZnO/MWCNTs as a catalytic degradable material against RhB.

4. Conclusions

In conclusion, we successfully prepared a new and sophisticated form of ZnO/MWCNT embedded with Ag nanoparticles. A dual procedure of pulsed laser ablation in liquid medium was successfully used to fabricate Ag-doped ZnO/MWCNTs, which might be used as an RhB-degradable catalytic material. ZnO/MWCNTs were created using the PLA of a Zn target immersed in f-MWCNTs, which served as a solvent and matrix source for a further PLA of Ag immersed in the ZnO/MWCNTs solution that was created in the first PLA process. Notably, no extra reducing agents or surfactants were required in the synthesis. To characterize the produced materials, XRD, UV-vis, FE-SEM, XPS, and EDX were used. After that, the adsorption study showed that the effective RhB degradation from aqueous solutions was accomplished by the adsorbent, with an even better removal at low concentrations of about 40 mg/L of the adsorbent. According to the research, the adsorption is pH dependent, with pH 11 producing the greatest clearance. The removal process was controlled using pseudo-first-order rate kinetics. According to the adsorption results, RhB may be effectively removed from aqueous solutions using a hybrid composed of Ag-doped ZnO/MWCNTs.

Author Contributions: Conceptualization, A.M.M., T.A.A. and E.A.M.; methodology, S.M.A., R.A.R., E.A.M. and F.H.A.; software, H.A.A., T.A.A., F.H.A., R.A.R. and A.M.M.; formal analysis, A.M.M. and H.A.A.; investigation, A.M.M., R.A.R., R.A.P., S.M.A. and E.A.M.; resources, T.A.A., E.A.M. and F.H.A.; data curation, H.A.A.; writing—original draft preparation, A.M.M., E.A.M. and H.A.A.; writing—review and editing, F.H.A., R.A.R. and R.A.P.; visualization, A.M.M., E.A.M. and R.A.R.; supervision, A.M.M.; project administration, A.M.M., T.A.A. and F.H.A.; funding acquisition, F.H.A. All authors have read and agreed to the published version of the manuscript.

Funding: This research was funded by Princess Nourah bint Abdulrahman University Researchers Supporting Project number (PNURSP2022R71), Princess Nourah bint Abdulrahman University, Riyadh, Saudi Arabia.

Institutional Review Board Statement: Not applicable.

Informed Consent Statement: Not applicable.

Data Availability Statement: Not applicable.

Acknowledgments: The authors express their gratitude to Princess Nourah bint Abdulrahman University Researchers Supporting Project number (PNURSP2022R71), Princess Nourah bint Abdulrahman University, Riyadh, Saudi Arabia. Additionally, the authors would like to thank the Deanship of Scientific Research at Umm Al-Qura University for supporting this work by Grant Code: (22UQU4320141DSR60).

Conflicts of Interest: The authors declare no conflict of interest.

References

1. Velidandi, A.; Pabbathi, N.P.P.; Baadhe, R.R. Study of parameters affecting the degradation of rhodamine-B and methyl orange dyes by Annona muricata leaf extract synthesized nanoparticles as well as their recyclability. *J. Mol. Struct.* **2021**, *1236*, 130287. [CrossRef]
2. Hu, C.; Le, A.T.; Pung, S.Y.; Stevens, L.; Neate, N.; Hou, X.; Grant, D.; Xu, F. Efficient dye-removal via Ni-decorated graphene oxide-carbon nanotube nanocomposites. *Mater. Chem. Phys.* **2021**, *260*, 124117. [CrossRef]
3. Wang, Y.; Jiang, F.; Chen, J.; Sun, X.; Xian, T.; Yang, H. In Situ Construction of CNT/CuS Hybrids and Their Application in Photodegradation for Removing Organic Dyes. *Nanomaterials* **2020**, *10*, 178. [CrossRef] [PubMed]
4. Li, Y.; Wang, Y.; Lu, H.; Li, X. Preparation of $CoFe_2O_4$–P4VP@Ag NPs as effective and recyclable catalysts for the degradation of organic pollutants with NaBH4 in water. *Int. J. Hydrogen Energy* **2020**, *45*, 16080–16093. [CrossRef]
5. Chen, Y.; Qian, J.; Wang, N.; Xing, J.; Liu, L. In-situ synthesis of CNT/TiO_2 heterojunction nanocomposite and its efficient photocatalytic degradation of Rhodamine B dye. *Inorg. Chem. Commun.* **2020**, *119*, 108071. [CrossRef]
6. Ait-Touchente, Z.; Khalil, A.M.; Simsek, S.; Boufi, S.; Ferreira, L.F.V.; Vilar, M.R.; Touzani, R.; Chehimi, M.M. Ultrasonic effect on the photocatalytic degradation of Rhodamine 6G (Rh6G) dye by cotton fabrics loaded with TiO_2. *Cellulose* **2020**, *27*, 1085–1097. [CrossRef]
7. Li, Y.; Lu, H.; Wang, Y.; Li, X. Deposition of Au nanoparticles on PDA-functionalized PVA beads as a recyclable catalyst for degradation of organic pollutants with NaBH4 in aqueous solution. *J. Alloys Compd.* **2019**, *793*, 115–126. [CrossRef]
8. Gai, G.; Wang, L.; Zhao, L.; Bi, F.; Xiao, S.; Zhao, G.; Zou, S. Facile electrospinning fabrication of nickel oxide nanotubes and their photocatalytic properties. *J. Mater. Sci. Mater. Electron.* **2017**, *28*, 7271–7276. [CrossRef]
9. Mohamed, M.M.; Osman, G.; Khairou, K. Fabrication of Ag nanoparticles modified TiO_2–CNT heterostructures for enhanced visible light photocatalytic degradation of organic pollutants and bacteria. *J. Environ. Chem. Eng.* **2015**, *3*, 1847–1859. [CrossRef]
10. Guo, H.; Jiao, T.; Zhang, Q.; Guo, W.; Peng, Q.; Yan, X. Preparation of Graphene Oxide-Based Hydrogels as Efficient Dye Adsorbents for Wastewater Treatment. *Nanoscale Res. Lett.* **2015**, *10*, 272. [CrossRef]
11. Yan, Y.; Sun, H.; Yao, P.; Kang, S.-Z.; Mu, J. Effect of multi-walled carbon nanotubes loaded with Ag nanoparticles on the photocatalytic degradation of rhodamine B under visible light irradiation. *Appl. Surf. Sci.* **2011**, *257*, 3620–3626. [CrossRef]
12. Chen, M.-L.; Oh, W.-C. Photodegradation of organic dyes over nickel distributed CNT/TiO_2 composite synthesized by a simple sol-gel method. *Mater. Sci.* **2011**, *29*, 112–120. [CrossRef]

13. Mostafa, A.M.; Mwafy, E.A. Laser-assisted for preparation Ag/CdO nanocomposite thin film: Structural and optical study. *Opt. Mater.* **2020**, *107*, 110124. [CrossRef]
14. Mostafa, A.M.; Mwafy, E.A. Synthesis of ZnO/CdO thin film for catalytic degradation of 4-nitrophenol. *J. Mol. Struct.* **2020**, *1221*, 128872. [CrossRef]
15. Alkallas, F.H.; Ahmed, H.A.; Alrebdi, T.A.; Pashameah, R.A.; Alrefaee, S.H.; Alsubhe, E.; Trabelsi, A.B.G.; Mostafa, A.M.; Mwafy, E.A. Removal of Ni(II) Ions by Poly(Vinyl Alcohol)/Al_2O_3 Nanocomposite Film via Laser Ablation in Liquid. *Membranes* **2022**, *12*, 660. [CrossRef]
16. Mostafa, A.M.; Mwafy, E.A.; Awwad, N.S.; Ibrahium, H.A. Synthesis of multi-walled carbon nanotubes decorated with silver metallic nanoparticles as a catalytic degradable material via pulsed laser ablation in liquid media. *Colloids Surf. A Physicochem. Eng. Asp.* **2021**, *626*, 126992. [CrossRef]
17. Mostafa, A.M.; Mwafy, E.A.; Awwad, N.S.; Ibrahium, H.A. Linear and nonlinear optical studies of Ag/Zn/ZnO nanocomposite thin film prepared by pulsed laser deposition technique. *Radiat. Phys. Chem.* **2021**, *179*, 109233. [CrossRef]
18. Mostafa, A.M.; Mwafy, E.A.; Toghan, A. ZnO nanoparticles decorated carbon nanotubes via pulsed laser ablation method for degradation of methylene blue dyes. *Colloids Surf. A Physicochem. Eng. Asp.* **2021**, *627*, 127204. [CrossRef]
19. Mostafa, A.M.; Menazea, A. Laser-assisted for preparation ZnO/CdO thin film prepared by pulsed laser deposition for catalytic degradation. *Radiat. Phys. Chem.* **2020**, *176*, 109020. [CrossRef]
20. Alkallas, F.H.; Ahmed, H.A.; Pashameah, R.A.; Alrefaee, S.H.; Toghan, A.; Trabelsi, A.B.G.; Mostafa, A.M. Nonlinearity enhancement of Multi-walled carbon nanotube decorated with ZnO nanoparticles prepared by laser assisted method. *Opt. Laser Technol.* **2022**, *155*, 108444. [CrossRef]
21. ElFaham, M.M.; Mostafa, A.M.; Mwafy, E.A. The effect of reaction temperature on structural, optical and electrical properties of tunable ZnO nanoparticles synthesized by hydrothermal method. *J Phys Chem Solids.* **2021**, *154*, 110089. [CrossRef]
22. Mostafa, A.M.; Mwafy, E.A.; Lotfy, V.F.; Basta, A.H. Optical, electrical and mechanical studies of paper sheets coated by metals (Cu and Ag) via pulsed laser deposition. *J. Mol. Struct.* **2019**, *1198*, 126927. [CrossRef]
23. Mostafa, A.M.; Lotfy, V.F.; Mwafy, E.A.; Basta, A.H. Influence of coating by Cu and Ag nanoparticles via pulsed laser deposition technique on optical, electrical and mechanical properties of cellulose paper. *J. Mol. Struct.* **2020**, *1203*, 127472. [CrossRef]
24. Mostafa, A.M.; Mwafy, E.A.; Awwad, N.S.; Ibrahium, H.A. Au@Ag core/shell nanoparticles prepared by laser-assisted method for optical limiting applications. *J. Mater. Sci. Mater. Electron.* **2021**, *32*, 14728–14739. [CrossRef]
25. Alkallas, F.H.; Toghan, A.; Ahmed, H.A.; Alrefaee, S.H.; Pashameah, R.A.; Alrebdi, T.A.; Mwafy, E.A.; Mostafa, A.M. Catalytic performance of NiO nanoparticles decorated carbon nanotubes via one-pot laser ablation method against methyl orange dye. *J. Mater. Res. Technol.* **2022**, *18*, 3336–3346. [CrossRef]
26. Altowyan, A.S.; Toghan, A.; Ahmed, H.A.; Pashameah, R.A.; Mwafy, E.A.; Alrefaee, S.H.; Mostafa, A.M. Removal of methylene blue dye from aqueous solution using carbon nanotubes decorated by nickel oxide nanoparticles via pulsed laser ablation method. *Radiat. Phys. Chem.* **2022**, *198*, 110268. [CrossRef]
27. Mwafy, E.A.; Mostafa, A.M.; Awwad, N.S.; Ibrahium, H.A. Catalytic activity of multi-walled carbon nanotubes decorated with tungsten trioxides nanoparticles against 4-nitrophenol. *J. Phys. Chem. Solids* **2021**, *158*, 110258. [CrossRef]
28. Abozied, A.M.; Mostafa, A.M.; Abouelsayed, A.; Hassan, A.F.; Ramadan, A.A.; Al-Ashkar, E.A.; Anis, A. Preparation, characterization, and nonlinear optical properties of graphene oxide thin film doped with low chirality metallic SWCNTs. *J. Mater. Res. Technol.* **2021**, *12*, 1461–1472. [CrossRef]
29. Alrebdi, T.A.; Ahmed, H.A.; Alkallas, F.H.; Mwafy, E.A.; Trabelsi, A.B.G.; Mostafa, A.M. Structural, linear and nonlinear optical properties of NiO nanoparticles–multi-walled carbon nanotubes nanocomposite for optoelectronic applications. *Radiat. Phys. Chem.* **2022**, *195*, 110088. [CrossRef]
30. Alamro, F.S.; Mostafa, A.M.; Ahmed, H.A.; Toghan, A. Zinc oxide/carbon nanotubes nanocomposite: Synthesis, characterization and catalytic reduction of 4-nitrophenol via laser assistant method. *Surf. Interfaces.* **2021**, *26*, 101406. [CrossRef]
31. Mostafa, A.M. The enhancement of nonlinear absorption of Zn/ZnO thin film by creation oxygen vacancies via infrared laser irradiation and coating with Ag thin film via pulsed laser deposition. *J. Mol. Struct.* **2021**, *1226*, 129407. [CrossRef]
32. Mostafa, A.M. Preparation and study of nonlinear response of embedding ZnO nanoparticles in PVA thin film by pulsed laser ablation. *J. Mol. Struct.* **2021**, *1223*, 129007. [CrossRef]
33. Altowyan, A.S.; Mostafa, A.M.; Ahmed, H.A. Effect of liquid media and laser energy on the preparation of Ag nanoparticles and their nanocomposites with Au nanoparticles via laser ablation for optoelectronic applications. *Optik* **2021**, *241*, 167217. [CrossRef]
34. Mostafa, A.M.; Mwafy, E.A.; Awwad, N.S.; Ibrahium, H.A. Catalytic activity of Ag nanoparticles and Au/Ag nanocomposite prepared by pulsed laser ablation technique against 4-nitrophenol for environmental applications. *J. Mater. Sci. Mater. Electron.* **2021**, *32*, 11978–11988. [CrossRef]
35. ElFaham, M.M.; Okil, M.; Mostafa, A.M. Effects of post-laser irradiation on the optical and structure properties of Al_2O_3 nanoparticles produced by laser ablation. *J. Appl. Phys.* **2020**, *128*, 153104. [CrossRef]
36. Al-Kadhi, N.S.; Pashameah, R.A.; Ahmed, H.A.; Alrefaee, S.H.; Alamro, F.S.; Faqih, H.H.; Mwafy, E.A.; Mostafa, A.M. Preparation of NiO/MWCNTs nanocomposite and its application for cadmium ion removal from aqueous solutions. *J. Mater. Res. Technol.* **2022**, *19*, 1961–1971. [CrossRef]

37. Alrebdi, T.A.; Ahmed, H.A.; Alsubhe, E.; Alkallas, F.A.; Mwafy, E.A.; Pashameah, R.A.; Toghan, A.; Mostafa, A.M. Synthesis of NiO-PVA nanocomposite by laser assisted-method and its characterization as a novel adsorbent for removal phosphate from aqueous water. *Opt. Laser Technol.* **2022**, *156*, 108526. [CrossRef]
38. Alamro, F.S.; Toghan, A.; Ahmed, H.A.; Mostafa, A.M.; Alakhras, A.I.; Mwafy, E.A. Multifunctional leather surface embedded with zinc oxide nanoparticles by pulsed laser ablation method. *Microsc. Res. Tech.* **2022**, *85*, 4, 1611–1617.
39. Mwafy, E.A.; Mostafa, A.M. Efficient removal of Cu (II) by SnO_2/MWCNTs nanocomposite by pulsed laser ablation method. *Nano-Struct. Nano Objects* **2020**, *24*, 100591. [CrossRef]
40. Mwafy, E.A.; Mostafa, A.M. Multi walled carbon nanotube decorated cadmium oxide nanoparticles via pulsed laser ablation in liquid media. *Opt. Laser Technol.* **2019**, *111*, 249–254. [CrossRef]
41. Alamro, F.S.; Mostafa, A.M.; Abu Al-Ola, K.A.; Ahmed, H.A.; Toghan, A. Synthesis of Ag Nanoparticles-Decorated CNTs via Laser Ablation Method for the Enhancement the Photocatalytic Removal of Naphthalene from Water. *Nanomaterials* **2021**, *11*, 2142. [CrossRef]
42. Liu, Z.; Zhang, W.; Liang, Q.; Huang, J.; Shao, B.; Liu, Y.; Liu, Y.; He, Q.; Wu, T.; Gong, J.; et al. Microwave-assisted high-efficiency degradation of methyl orange by using $CuFe_2O_4$/CNT catalysts and insight into degradation mechanism. *Environ. Sci. Pollut. Res.* **2021**, *28*, 42683–42693. [CrossRef] [PubMed]
43. Kamali, N.; Mehrabadi, A.R.; Mirabi, M.; Zahed, M.A. Comparison of micro and nano MgO-functionalized vinasse biochar in phosphate removal: Micro-nano particle development, RSM optimization, and potential fertilizer. *J. Water Process Eng.* **2021**, *39*, 101741. [CrossRef]
44. Huo, J.-B.; Yu, G.; Wang, J. Adsorptive removal of Sr(II) from aqueous solution by polyvinyl alcohol/graphene oxide aerogel. *Chemosphere* **2021**, *278*, 130492. [CrossRef] [PubMed]
45. Howe, J.Y.; Rawn, C.J.; Jones, L.E.; Ow, H. Improved crystallographic data for graphite. *Powder Diffr.* **2003**, *18*, 150–154. [CrossRef]
46. Zhang, J.; Sun, L.; Liao, C.; Yan, C. A simple route towards tubular ZnO. *Chem. Commun.* **2002**, *3*, 262–263. [CrossRef] [PubMed]
47. Kalaivani, R.; Maruthupandy, M.; Muneeswaran, T.; Beevi, A.H.; Anand, M.; Ramakritinan, C.; Kumaraguru, A. Synthesis of chitosan mediated silver nanoparticles (Ag NPs) for potential antimicrobial applications. *Front. Lab. Med.* **2018**, *2*, 30–35. [CrossRef]
48. Li, M.; Huang, W.; Qian, W.; Liu, B.; Lin, H.; Li, W.; Wan, L.; Dong, C. Controllable Ag nanoparticle coated ZnO nanorod arrays on an alloy substrate with enhanced field emission performance. *RSC Adv.* **2017**, *7*, 46760–46766. [CrossRef]
49. Li, Z.; Yin, L. Sandwich-like reduced graphene oxide wrapped MOF-derived $ZnCo_2O_4$–ZnO–C on nickel foam as anodes for high performance lithium ion batteries. *J. Mater. Chem. A* **2015**, *3*, 21569–21577. [CrossRef]
50. Zhao, F.; Qian, W.; Li, M.; Li, W.; Chen, L.; Zhong, F.; Huang, W.; Dong, C. Directly grown carbon nanotube based hybrid electrodes with enhanced thermo-cell performances. *RSC Adv.* **2017**, *7*, 23890–23895. [CrossRef]
51. Yang, Z.-F.; Li, L.-Y.; Hsieh, C.-T.; Juang, R.-S. Co-precipitation of magnetic Fe3O4 nanoparticles onto carbon nanotubes for removal of copper ions from aqueous solution. *J. Taiwan Inst. Chem. Eng.* **2018**, *82*, 56–63. [CrossRef]
52. Kim, J.D.; Yun, H.; Kim, G.C.; Lee, C.W.; Choi, H.C. Antibacterial activity and reusability of CNT-Ag and GO-Ag nanocomposites. *Appl. Surf. Sci.* **2013**, *283*, 227–233. [CrossRef]
53. Mwafy, E.A.; Gaafar, M.; Mostafa, A.M.; Marzouk, S.; Mahmoud, I. Novel laser-assisted method for synthesis of SnO_2/MWCNTs nanocomposite for water treatment from Cu (II). *Diam. Relat. Mater.* **2021**, *113*, 108287. [CrossRef]
54. Saravanakkumar, D.; Devi, S.U.; Sivaranjani, S.; Gnanasaravanan, S. Nano engineering, Applications, Structural investigation on synthesized Ag doped ZnO-MWCNT and Its applications. *J. Nano Sci. Nano Eng. Appl.* **2018**, *8*, 2321–5194.
55. Yousif, L.N.; Ibrahim, N.M.; Kamel, R.I.; Rahmah, M.I. Synthesis And Studied Structural and Morphological Properties of 1-Dimensional Zno-Ag_2O-Ag Nanowire. *IOP Conf. Ser. Earth Environ. Sci.* **2022**, *961*, 012016. [CrossRef]
56. Basha, M.A.-F.; Mostafa, A.M. UV-induced macromolecular and optical modifications in gelatin solid films with transition metal chlorides. *J. Mol. Struct.* **2019**, *1182*, 181–190. [CrossRef]
57. Chan, Y.Y.; Pang, Y.L.; Lim, S.; Lai, C.W.; Abdullah, A.Z.; Chong, W.C. Biosynthesized Fe- and Ag-doped ZnO nanoparticles using aqueous extract of *Clitoria ternatea* Linn for enhancement of sonocatalytic degradation of Congo red. *Environ. Sci. Pollut. Res.* **2020**, *27*, 34675–34691. [CrossRef]
58. Zhu, J.; Zhang, X.; Qin, Z.; Zhang, L.; Ye, Y.; Cao, M.; Gao, L.; Jiao, T. Preparation of PdNPs doped chitosan-based composite hydrogels as highly efficient catalysts for reduction of 4-nitrophenol. *Colloids Surf. A Physicochem. Eng. Asp.* **2021**, *611*, 125889. [CrossRef]
59. Sharma, K.; Majhi, S.; Ali, M.; Singh, R.; Tripathi, C.S.P.; Guin, D. Fabrication of Reduced Graphene Oxide-Silver/Polyvinyl Alcohol Nanocomposite Film for Reduction of 4-Nitrophenol and Methyl Orange Dye. *ChemistrySelect* **2021**, *6*, 6071–6076. [CrossRef]
60. Cai, Y.; Yang, F.; Wu, L.; Shu, Y.; Qu, G.; Fakhri, A.; Gupta, V.K. Hydrothermal-ultrasonic synthesis of CuO nanorods and CuWO4 nanoparticles for catalytic reduction, photocatalysis activity, and antibacterial properties. *Mater. Chem. Phys.* **2021**, *258*, 123919. [CrossRef]
61. Yang, T.; Tang, Y.; Liu, L.; Gao, Y.; Zhang, Y. Cu-anchored CNTs for effectively catalytic reduction of 4-nitrophenol. *Chem. Phys.* **2020**, *533*, 110738. [CrossRef]
62. You-Ji, L.; Wei, C. Photocatalytic degradation of Rhodamine B using nanocrystalline TiO_2–zeolite surface composite catalysts: Effects of photocatalytic condition on degradation efficiency. *Catal. Sci. Technol.* **2011**, *1*, 802–809. [CrossRef]

63. Byrappa, K.; Subramani, A.K.; Ananda, S.; Rai, K.M.L.; Dinesh, R.; Yoshimura, M. Photocatalytic degradation of rhodamine B dye using hydrothermally synthesized ZnO. *Bull. Mater. Sci.* **2006**, *29*, 433–438. [CrossRef]
64. Le, A.T.; Samsuddin, N.S.B.; Chiam, S.-L.; Pung, S.-Y. Synergistic effect of pH solution and photocorrosion of ZnO particles on the photocatalytic degradation of Rhodamine B. *Bull. Mater. Sci.* **2021**, *44*, 5. [CrossRef]
65. Zulfiqar, M.; Lee, S.Y.; Mafize, A.A.; Kahar, N.A.M.A.; Johari, K.; Rabat, N.E. Efficient Removal of Pb(II) from Aqueous Solutions by Using Oil Palm Bio-Waste/MWCNTs Reinforced PVA Hydrogel Composites: Kinetic, Isotherm and Thermodynamic Modeling. *Polymers* **2020**, *12*, 430. [CrossRef]
66. Shi, Y.; Xing, Y.; Deng, S.; Zhao, B.; Fu, Y.; Liu, Z. Synthesis of proanthocyanidins-functionalized Fe_3O_4 magnetic nanoparticles with high solubility for removal of heavy-metal ions. *Chem. Phys. Lett.* **2020**, *753*, 137600. [CrossRef]
67. Mukwevho, N.; Gusain, R.; Fosso-Kankeu, E.; Kumar, N.; Waanders, F.; Ray, S.S. Removal of naphthalene from simulated wastewater through adsorption-photodegradation by ZnO/Ag/GO nanocomposite. *J. Ind. Eng. Chem.* **2020**, *81*, 393–404. [CrossRef]
68. Sadegh, H.; Ali, G.A.; Makhlouf, A.S.H.; Chong, K.F.; Alharbi, N.S.; Agarwal, S.; Gupta, V.K. MWCNTs-Fe_3O_4 nanocomposite for Hg(II) high adsorption efficiency. *J. Mol. Liq.* **2018**, *258*, 345–353. [CrossRef]
69. Khan, M.M.R.; Akter, M.; Amin, K.; Younus, M.; Chakraborty, N. Synthesis, Luminescence and Thermal Properties of PVA–ZnO–Al_2O_3 Composite Films: Towards Fabrication of Sunlight-Induced Catalyst for Organic Dye Removal. *J. Polym. Environ.* **2018**, *26*, 3371–3381. [CrossRef]
70. Katheresan, V.; Kansedo, J.; Lau, S.Y. Efficiency of various recent wastewater dye removal methods: A review. *J. Environ. Chem. Eng.* **2018**, *6*, 4676–4697. [CrossRef]
71. Molla, A.; Sahu, M.; Hussain, S. Under dark and visible light: Fast degradation of methylene blue in the presence of Ag–In–Ni–S nanocomposites. *J. Mater. Chem. A* **2015**, *3*, 15616–15625. [CrossRef]
72. Hosseini, F.; Kasaeian, A.; Pourfayaz, F.; Sheikhpour, M.; Wen, D. Novel ZnO-Ag/MWCNT nanocomposite for the photocatalytic degradation of phenol. *Mater. Sci. Semicond. Process.* **2018**, *83*, 175–185. [CrossRef]
73. Rohilla, S.; Gupta, A.; Kumar, V.; Kumari, S.; Petru, M.; Amor, N.; Noman, M.T.; Dalal, J. Excellent UV-Light Triggered Photocatalytic Performance of ZnO.SiO_2 Nanocomposite for Water Pollutant Compound Methyl Orange Dye. *Nanomaterials* **2021**, *11*, 2548. [CrossRef] [PubMed]
74. Saravanakkumar, D.; Oualid, H.A.; Brahmi, Y.; Ayeshamariam, A.; Karunanaithy, M.; Saleem, A.M.; Kaviyarasu, K.; Sivaranjani, S.; Jayachandran, M. Synthesis and characterization of CuO/ZnO/CNTs thin films on copper substrate and its photocatalytic applications. *OpenNano* **2019**, *4*, 100025. [CrossRef]
75. Kumar, N.; Mittal, H.; Alhassan, S.; Ray, S.S. Bionanocomposite Hydrogel for the Adsorption of Dye and Reusability of Generated Waste for the Photodegradation of Ciprofloxacin: A Demonstration of the Circularity Concept for Water Purification. *ACS Sustain. Chem. Eng.* **2018**, *6*, 17011–17025. [CrossRef]

membranes

MDPI

Article

Ag/ZnO Thin Film Nanocomposite Membrane Prepared by Laser-Assisted Method for Catalytic Degradation of 4-Nitrophenol

Tahani A. Alrebdi [1], Hoda A. Ahmed [2,3], Fatemah H. Alkallas [1], Rami Adel Pashameah [4], Salhah H. Alrefaee [3], Emaan Alsubhe [5], Ayman M. Mostafa [6,7,*] and Eman A. Mwafy [7,8]

1 Department of Physics, College of Science, Princess Nourah bint Abdulrahman University, P.O. Box 84428, Riyadh 11671, Saudi Arabia; taalrebdi@pnu.edu.sa (T.A.A.); fhalkallas@pnu.edu.sa (F.H.A.)
2 Department of Chemistry, Faculty of Science, Cairo University, Cairo 12613, Egypt; ahoda@sci.cu.edu.eg
3 Chemistry Department, College of Sciences, Taibah University, Yanbu 30799, Saudi Arabia; srfaay@taibahu.edu.sa
4 Department of Chemistry, Faculty of Applied Science, Umm Al-Qura University, Makkah 24230, Saudi Arabia; rapasha@uqu.edu.sa
5 Physics Department, Faculty of Science, Taibah University, Yanbu 30799, Saudi Arabia; esobhe@taibahu.edu.sa
6 Spectroscopy Department, Physics Research Institute, National Research Centre, 33 El Bohouth St. (Former El Tahrir St.), Dokki, Giza 12622, Egypt
7 Laser Technology Unit, Center of Excellent for Advanced Science, National Research Centre, 33 El Bohouth St. (Former El Tahrir St.), Dokki, Giza 12622, Egypt; emanmwafynrc@gmail.com
8 Physical Chemistry Department, Advanced Materials Technology and Mineral Resources Research Institute, National Research Centre, Giza 12622, Egypt
* Correspondence: aymanmdarwish@gmail.com

Abstract: Zinc oxide thin film (ZnO thin film) and a silver-doped zinc oxide nanocomposite thin film (Ag/ZnO thin film) were prepared by the technique of the pulsed laser deposition at 600 °C to be applicable as a portable catalytic material for the removal of 4-nitrophenol. The nanocomposite was prepared by making the deposition of the two targets (Zn and Ag), and it was analyzed by different techniques. According to the XRD pattern, the hexagonal wurtzite crystalline form of Ag-doped ZnO NPs suggested that the samples were polycrystalline. Additionally, the shifting of the diffraction peaks to the higher angles, which denotes that doping reduces the crystallite size, illustrated the typical effect of the dopant Ag nanostructure on the ZnO thin film, which has an ionic radius less than the host cation. From SEM images, Ag-doping drastically altered the morphological characteristics and reduced the aggregation. Additionally, its energy band gap decreased when Ag was incorporated. UV spectroscopy was then used to monitor the catalysis process, and Ag/ZnO thin films had a larger first-order rate constant of the catalytic reaction K than that of ZnO thin film. According to the catalytic experiment results, the Ag/ZnO thin film has remarkable potential for use in environmentally-favorable applications.

Keywords: PLD; thin film; ZnO; nanocomposite; catalytic degradation; membrane

1. Introduction

Thin films are more useful structures to study than their bulk counterparts because of their versatility and positive characteristics. Thin-film assembly of surface-supported metal–metal oxides is becoming more and more ubiquitous, and this is expected to lead to new scientific discoveries as well as expanded application opportunities [1–5]. Examples of applications include membrane-based chemical species separation, silicon integration for electronic detection of volatile compounds, and electrochemical catalysis. ZnO can be considered as a one of the most important materials as it can be used in several application domains. Additionally, ZnO has great piezoelectric properties and exceptional chemical stability due to its impressive semiconductor performance. Over the past few decades, ZnO thin films have been the subject of substantial research [5–8]. It has a number of clearly stated benefits, including being affordable, plentiful, safe to use, chemically stable, very transparent in the visible and near-infrared spectral area, ecologically benign, and highly catalytic. Since ZnO has a direct bandgap, it offers special features and the versatility to be employed in a variety of applications, including LEDs, gas sensors, optoelectronic components, UV lasers, solar cells, and screen displays [9–13].

The nanocomposite semiconductor thin film is formed by doping the semiconductor thin film with another thin film or metallic nanoparticles, which helps to enhance the optoelectronic properties. These materials provide ZnO with n-type conductivity because the presence of oxygen vacancies leads to the creation of Zn^{+2} interstitials. However, it can be doped to achieve P-type conductivity, making it promising for use in research. According to theoretical research, silver and gold are an appropriate dopant to obtain P-type ZnO. Ag is the most suitable choice among all of the transition metals due to its solubility and the fact that it contains one outermost electron, which accounts for its huge ionic size and orbital energy. In addition, compared to other transition metals, Ag is a cheap, non-toxic, superior electrical and high thermal conductor. The main reason for choosing ZnO doped with Ag is because it improves the characteristics and hinders the trapping of photo-generated electrons, which lengthens the lifespan of charge-separated states [14–18].

The pulsed laser deposition technique (PLD) has been found to be the most efficient and quick method for producing stoichiometric thin film by utilizing multiple targets; these are appealing qualities that make it possible to produce multi-component thin films of high quality. Thin films of metal oxide and other materials related to metal oxide are routinely produced using PLD. Metal nanoparticle-doped oxide films have recently been demonstrated to possess intriguing characteristics such as significant nonlinearity, photocatalysis, etc. [19–22].

Herein, we synthesized ZnO thin film and developed this structure by embedding the structure with Ag nanostructure to form Ag-doped ZnO nanocomposite thin films by PLD to apply in water treatment. The novelty in this work was based on filling vacancies in the zinc matrix and interstitial zinc atoms with Ag atoms. Different techniques were used to investigate the prepared thin films and determine the changes in their characteristics. Then, the prepared films were used as catalytic-degradable material for the water purification field.

2. Experimental Work

2.1. Materials

Merck supplied the 4-nitrophenol (4-NP), zinc dust target (Zn), silver target (Ag), sodium borohydride ($NaBH_4$), and mono-crystalline quartz substrate ($4 \times 1 \times 0.1$ cm^3), which was washed by chromic acid (purchased from Nasr, Egypt) and then thoroughly rinsed with distilled water, followed by sonication in an ultrasonic bath for 30 min. After treatment, the slides were stored in water until required for use.

2.2. Preparation Method

As previously reported, to adjust the optimum condition of dual-pulsed laser deposition [23–25], the thin film was created via PLD of the metal target at the following specifications: 600 °C reaction temperature, 7 ns pulse duration, 1064 nm wavelength, 30 min ablation time, and 7.5 W average power. As shown in Figure 1, pulsed laser ablation was first performed on the Zn to create ZnO thin films on quartz substrate, then on the Ag to create Ag/ZnO nanocomposite thin films on quartz substrate. For Ag-doped ZnO thin film, the film thickness was determined using a profilometer (SurfTestSJ-301) and was found to be 300 nm for ZnO thin film and 500 nm for Ag/ZnO thin film.

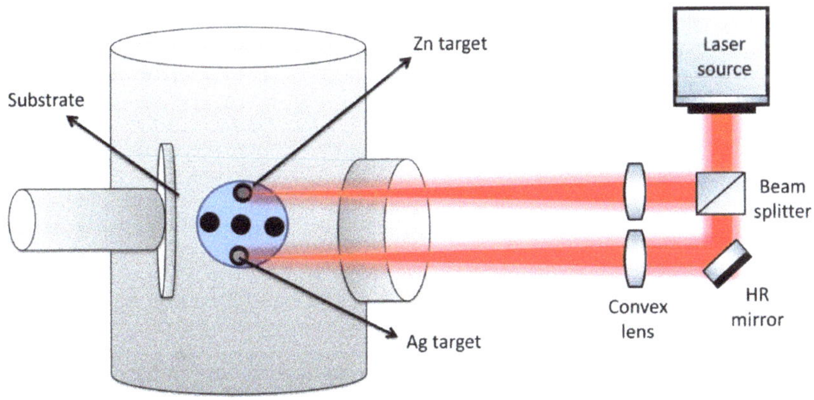

Figure 1. Representative graph of the preparation thin film of Ag/ZnO by PLD.

2.3. Characterization Techniques

UV-visible spectrophotometer (JASCO 570 UV–Vis, Japan), X-ray diffraction (XRD, Schimadzu 7000, Japan), scanning electron microscopic and energy dispersive X-ray (SEM-EDX, Quanta FEG 250, FEI, Czech Republic), and photoluminescence Spectrometer (PL, JASCO, FP-6500, Japan) were used for the characterization.

2.4. Catalytic Removal Procedure of 4-NP

The catalytic degradation procedure was conducted by adding the following items together: 100 μL of 0.1 M $NaBH_4$ (reducing agent), 2.77 mL of 0.4 mM 4-NP (pollutant) and the prepared thin films were used as a catalytic degradable material. Then, the UV-visible spectrum was examined every 10 min in order to examine the catalytic reaction.

3. Results

3.1. Investigation of the Prepared Films

Figure 2 shows the diffractogram of ZnO thin film and its embedding with Ag thin films to create a nanocomposite structure thin film. For ZnO thin film, a number of diffraction peaks that appeared corresponded to the reflection planes (1 0 0), (0 2 2), (1 0 2), (1 1 0), (1 0 3), (2 0 0), (1 1 2), and (2 0 1), which is an exact match to the XRD diffraction pattern of pure ZnO (JCPDS card No. 80-0075) [26]. There are no phases that matched any other oxides other than ZnO, which can be seen in the diffractogram for ZnO thin films. Additionally, the diffractogram of the prepared ZnO thin film has a prominent ZnO (0 0 2) peak, indicating that the films are oriented with the c-axis normal to the substrate with a high intensity peak and a sharp shape. In another words, for the ZnO thin film embedded with Ag nanostructure, three XRD peaks of the (1 1 1), (2 0 0), and (2 2 0) planes were indexed to silver, which was fcc phase-matched, confirming Ag as the secondary phase based on the JCPDS card No. 04-0783 [27]. Additionally, the Scherrer formula was used to determine the crystalline size for the (1 0 1) plane, the highest characteristic peak [28,29].

$$\mathbf{D} = \frac{(0.9)\boldsymbol{\lambda}}{\boldsymbol{\beta}\mathbf{cos}\boldsymbol{\theta}}$$

where λ is the wavelength of X-ray 1.54 Å, β is FWHM and θ is the diffraction angle of the (1 0 1) plane. The crystallite size decreases from 42 to 19 nm as the Ag dopant concentration increases. As the amount of Ag dopant in thin films increases, the lattice strain between crystallites also increases. More Ag ions cause the lattice to become disordered, which results in a reduction in crystallite size. Because Ag^+ ions have a higher radius (0.126 nm) than Zn^{2+} ions, the lattice constants rise, increasing the cell volume (0.074 nm).

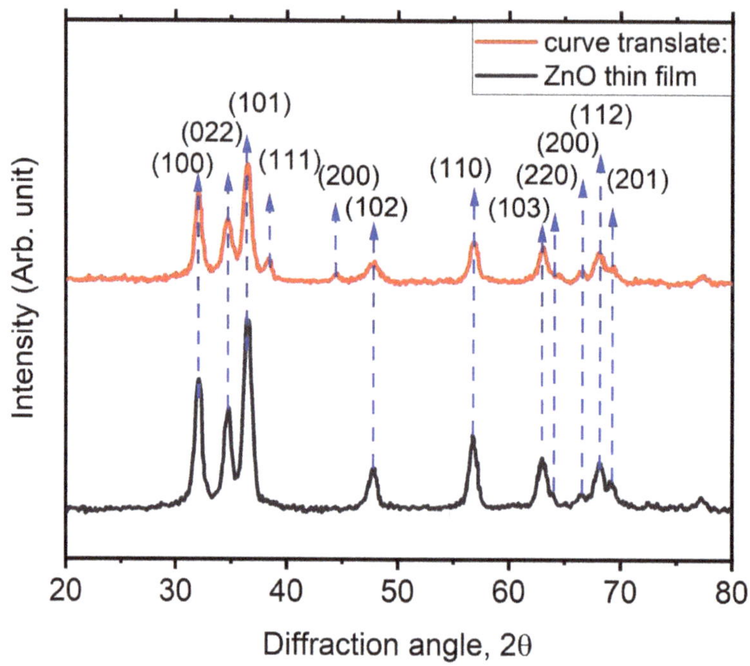

Figure 2. XRD diffractogram of ZnO thin film before and after being embedded with Ag thin films.

Figure 3 displays SEM images of the annealed ZnO and annealed ZnO/Ag films. The pulsed laser deposition of undoped ZnO films typically results in dense films. It was clear that in the case of Ag-doped films, bright spots started to appear, which is related to the embedding of the ZnO thin film with Ag film. Moreover, the ZnO/Ag thin films' elemental composition was examined using EDX spectra to establish the presence of Ag. A typical EDX result for thin films is shown in Figure 3. No other components were found, and the top concentrations were Zn, O, and Ag, which was confirmed by elemental mapping.

UV-vis spectrophotometer was used to evaluate the varying in the optical properties of the prepared strucuture. Figure 4 displays the UV-vis optical absorption spectra of ZnO and Ag/ZnO thin films. Strong excitonic peaks with significant absorbance strength are visible in the sample at 372 nm (pure ZnO). In addition, compared to pure ZnO, the absorption edge shifted to a longer wavelength with Ag deposition to create Ag/ZnO thin films. The inclusion of Ag into the ZnO matrix may be responsible for this shift in the absorption edge. Particles in doped ZnO may be smaller than the exciton Bohr radius, according to a shift in the absorption spectra [30–32]. This decrease in the transmittance values of the Ag/ZnO thin film in comparison to ZnO thin film may be caused by grain boundary scattering and the visible light absorption caused by surface plasmon resonance (SPR), which is in excellent agreement with previous studies in the literature [33,34].

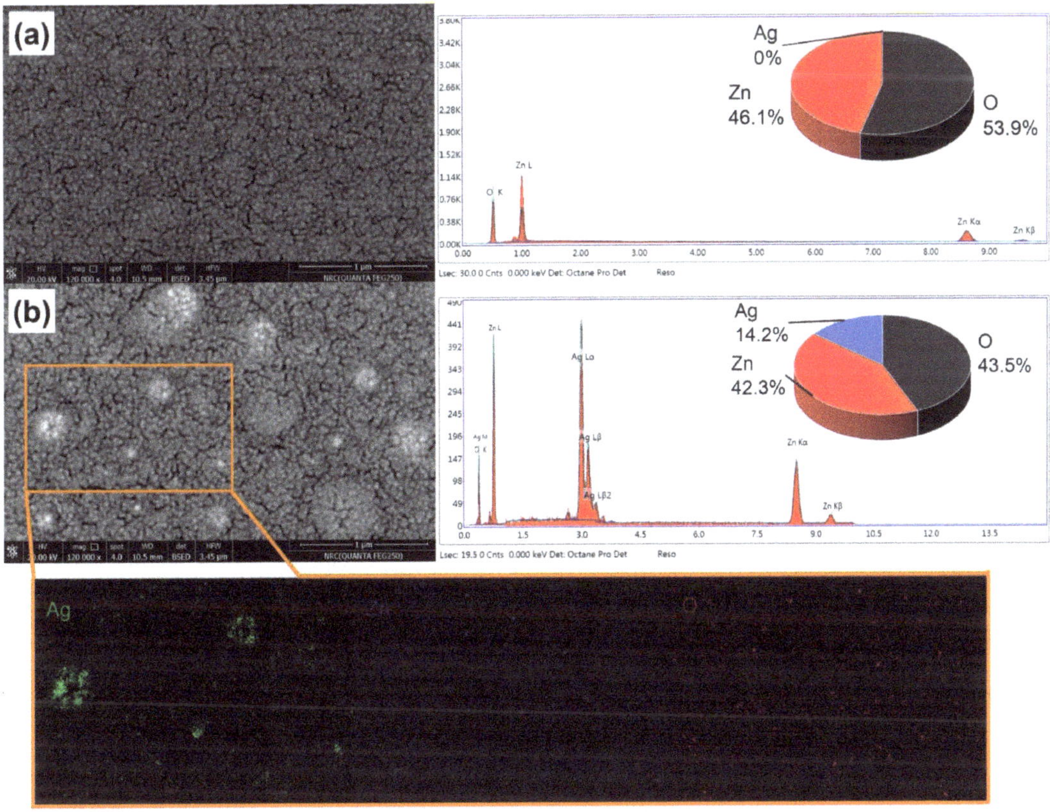

Figure 3. SEM image and elemental analysis of the prepared thin films of (**a**) ZnO and (**b**) Ag/ZnO.

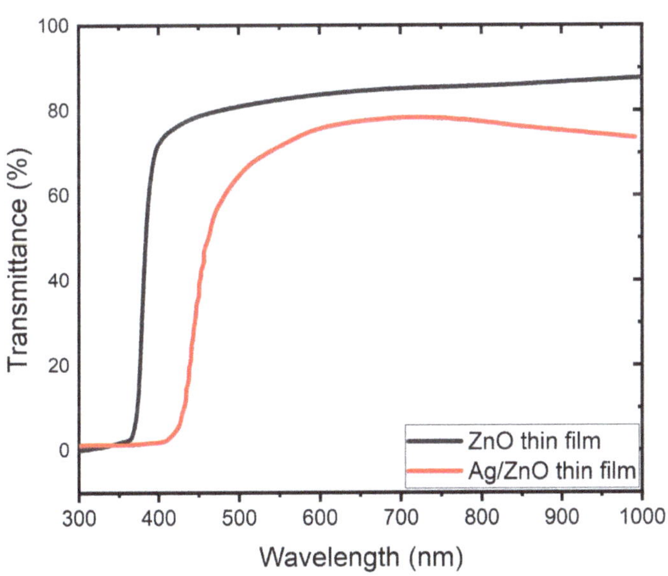

Figure 4. Absorbance spectra of ZnO thin film before and after being embedded with Ag thin films.

Figure 5 displays the PL spectra of thin films, which were calculated using an excitation wavelength of 325 nm and an emission spectrum that ranged from 360 to 410 nm. The samples' PL spectra show a single violet luminescence peak with a center wavelength of 378 nm. It has been reported that thin ZnO films produced on quartz via PLD emit violet fluorescence, which represents a mixture from the oxygen or zinc vacancies. In other words, in the case of Ag/ZnO nanocomposite thin film, the interface traps between Ag nanoclusters and ZnO grains that exist at the grain boundaries generate violet light from the electronic transition between CB and VB. Additionally, the emission spectra showed a substantial violet emission at 382 nm, which was probably generated by radiate defects brought on by interface traps at the grain boundaries. After Ag was added, this band edge emission peak migrated slightly to a higher value, indicating that the optical bandgap had shrunk. Besides, the presence of oxygen vacancies that interacted with Ag doping also had an impact on the location and intensity of this PL emission. Since Ag^{2+} ions take the position of Zn^{+} ions in the high concentration of Ag-doped ZnO, the peak's intensity is lower than it is for pure ZnO [35–38].

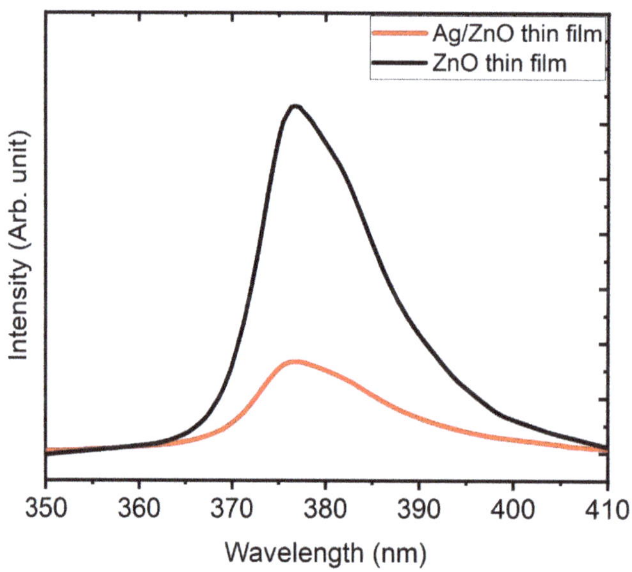

Figure 5. PL spectra of ZnO thin film before and after being embedded with Ag thin films.

3.2. Catalytic Activity

Then, using the UV-visible spectra in Figure 6, the catalyst's catalytic capabilities for reducing 4-NP in an excess of $NaBH_4$ were examined. The reduction of 4-NP in the presence of $NaBH_4$ at 298 K was catalyzed by the prepared thin films, as seen in the UV spectrum in Figure 6. The absorption of 4-NP was characterized via two peaks (314 nm and 400 nm) while the absorption characteristic peak of 4-NP ion was characterized by one peak (403 nm), and the absorption characteristic peak of the conversion from 4-NP to 4-AP was characterized by one peak (300 nm). This was related to four processes that took place in the catalytic reduction of 4-NP: (1) adsorption of the reactant 4-NP onto the thin ZnO or Ag/ZnO surface; (2) diffusion of 4-NP to the active site; (3) reduction on the site by $NaBH_4$; and (4) desorption of the product [39]. According to the UV-vis spectrum, ZnO thin films have a marginal catalytic effect in the first 20 min with an efficiency that does not exceed 35%, whereas Ag/ZnO thin films demonstrate excellent adsorption of 4-NP in the initial period of time with an efficiency that exceeds 90% [40–42].

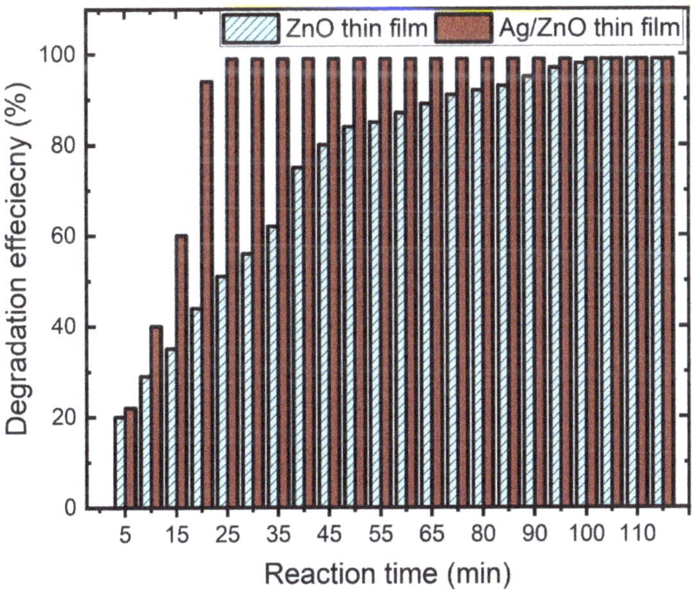

Figure 6. Degradation efficiency of ZnO thin film before and after being embedded with Ag thin films against 4-NP.

We inserted a single strip of the prepared thin film (1 cm × 2 cm) into a solution of deionized water containing 4-NP and $NaBH_4$ at 298 K in order to compare the catalytic effects of the ZnO and Ag/ZnO thin films. Figure 6 shows the degradation efficiency for ZnO thin film and Ag/ZnO thin film against the reaction process. As can be seen, the Ag/ZnO thin film completed the conversion of 4-NP to 4-AP in at least 27 min, whereas the ZnO thin film completed it in 105 min. The catalytic reaction took place after the catalyst accepted the electron donor BH^{4-}, transferred it to the electron acceptor 4-NP, and then allowed it to proceed. It might be argued that the rate of decrease was independent of the NaBH4 content because there was an overabundance of $NaBH_4$. The solution's color changed from light yellow-green to dark yellow-green when $NaBH_4$ was added to create a 4-NP ion.

According to Figure 7, which shows the relationship of $\ln(A_t/A_o)$ versus time (t), the reaction showed a pseudo first-order reaction based on their linear correlation, where A_t and A_o are the absorbance values at the times (any time, zero time), respectively. The pseudo 1st rate constant of the Ag/ZnO thin film was significantly greater than that of the ZnO thin film. One of the reasons was that the Ag/ZnO structures produced better mass transfer than ZnO structure-assisted catalysts.

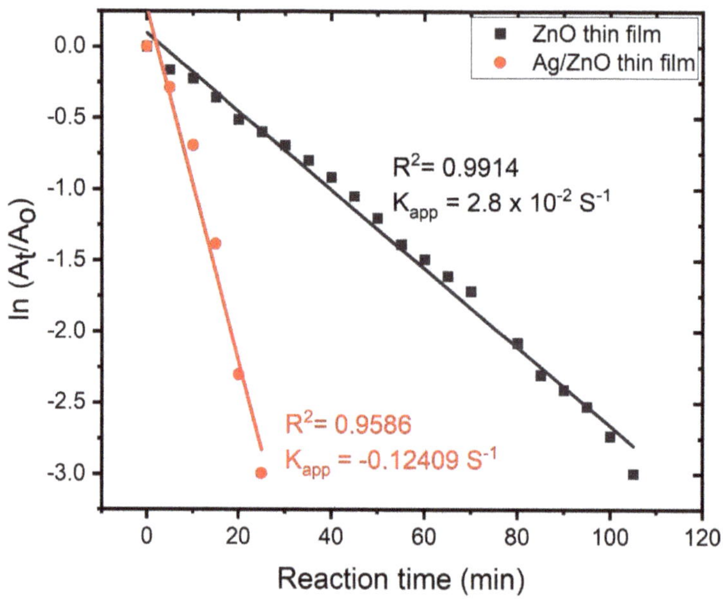

Figure 7. Investigation of $\ln(A_t/A_o)$ with respect to reaction time (t) for the thin films (ZnO and Ag/ZnO) against 4-NP.

4. Conclusions

In conclusion, Ag-nanocluster-doped ZnO films were produced at 600 °C by PLD. Several analytical techniques were utilized to examine the produced thin films' physico-chemical characteristics in order to learn more about their optical, structural, and morphological characteristics. These methods revealed the crystalline structure, excellent surface homogeneity, and the production of nano-structural particle size in both the produced films. Then, ZnO structure and the Ag/ZnO structure were investigated for the removal of 4-NP, it was discovered that the degradation efficiency of the Ag/ZnO thin film completed the conversion to 4-AP in at least 27 min, whereas the ZnO thin film competed it in 105 min.

Author Contributions: Conceptualization, T.A.A.; methodology, S.H.A., A.M.M., E.A., T.A.A., E.A.M. and F.H.A.; formal analysis, A.M.M., E.A.M. and H.A.A.; investigation, A.M.M., S.H.A., E.A. and E.A.M.; resources, T.A.A., A.M.M., R.A.P. and F.H.A.; data curation, H.A.A.; writing—original draft preparation, R.A.P. and H.A.A.; writing—review and editing, E.A.M.; supervision, A.M.M.; project administration, A.M.M. and T.A.A.; funding acquisition, T.A.A. All authors have read and agreed to the published version of the manuscript.

Funding: This research was funded by the Princess Nourah bint Abdulrahman University Researchers Supporting Project number (PNURSP2022R71), Princess Nourah bint Abdulrahman University, Riyadh, Saudi Arabia.

Acknowledgments: The authors extend their sincere appreciation to Princess Nourah bint Abdulrahman University Researchers Supporting Project number (PNURSP2022R71), Princess Nourah bint Abdulrahman University, Riyadh, Saudi Arabia. The authors would also like to thank the Deanship of Scientific Research at Umm Al-Qura University for supporting this work with Grant Code: (22UQU4320141DSR37).

Conflicts of Interest: The authors declare no conflict of interest.

References

1. Ahmad Fahri, M.A.S.; Rohizat, N.S.; Yudiana, A.; Fauzia, V.; Abdul Khudus, M.I.M.; Zakaria, R. Nonlinear optical response of zinc oxide nanorods with gold and silver alloy embedment. *Phys. B Condens. Matter* **2022**, *637*, 413899. [CrossRef]
2. Soudi, J.; Sandeep, K.M.; Sarojini, B.K.; Patil, P.S.; Maidur, S.R.; Balakrishna, K.M. Thermo-optic effects mediated self focusing mechanism and optical power limiting studies of ZnO thin films deposited on ITO coated PET substrates by RF magnetron sputtering under continuous wave laser regime. *Optik* **2021**, *225*, 165835. [CrossRef]
3. Seol, W.; Anoop, G.; Park, H.; Shin, C.W.; Lee, J.Y.; Kim, T.Y.; Kim, W.S.; Joh, H.; Samanta, S.; Jo, J.Y. Ferroelectricity in solution-processed V-doped ZnO thin films. *J. Alloys Compd.* **2021**, *853*, 157369. [CrossRef]
4. Mostafa, A.M. The enhancement of nonlinear absorption of Zn/ZnO thin film by creation oxygen vacancies via infrared laser irradiation and coating with Ag thin film via pulsed laser deposition. *J. Mol. Struct.* **2021**, *1226*, 129407. [CrossRef]
5. Mostafa, A.M. Preparation and study of nonlinear response of embedding ZnO nanoparticles in PVA thin film by pulsed laser ablation. *J. Mol. Struct.* **2021**, *1223*, 129007. [CrossRef]
6. Alkallas, F.H.; Ahmed, H.A.; Pashameah, R.A.; Alrefaee, S.H.; Toghan, A.; Trabelsi, A.B.G.; Mostafa, A.M. Nonlinearity enhancement of Multi-walled carbon nanotube decorated with ZnO nanoparticles prepared by laser assisted method. *Opt. Laser Technol.* **2022**, *155*, 108444. [CrossRef]
7. ElFaham, M.M.; Mostafa, A.M.; Mwafy, E.A. The effect of reaction temperature on structural, optical and electrical properties of tunable ZnO nanoparticles synthesized by hydrothermal method. *J .Phys. Chem. Solids* **2021**, *154*, 110089. [CrossRef]
8. Upadhya, K.; Deekshitha, U.G.; Antony, A.; Ani, A.; Kityk, I.V.; Jedryka, J.; Wojciechowski, A.; Ozga, K.; Poornesh, P.; Kulkarni, S.D.; et al. Second and third harmonic nonlinear optical process in spray pyrolysed Mg: ZnO thin films. *Opt. Mater.* **2020**, *102*, 109814. [CrossRef]
9. Mostafa, A.M.; Mwafy, E.A.; Toghan, A. ZnO nanoparticles decorated carbon nanotubes via pulsed laser ablation method for degradation of methylene blue dyes. *Colloids Surf. A Physicochem. Eng. Asp.* **2021**, *627*, 127204. [CrossRef]
10. Alamro, F.S.; Toghan, A.; Ahmed, H.A.; Mostafa, A.M.; Alakhras, A.I.; Mwafy, E.A. Multifunctional leather surface embedded with zinc oxide nanoparticles by pulsed laser ablation method. *Microsc. Res. Tech.* **2022**, *85*, 1611–1617. [CrossRef]
11. Alturki, A.M. Effect of Preparation Method on the Particles Size, Dielectric Constant and Antibacterial Properties of ZnO Nanoparticles and Thin Film of ZnO/Chitosan. *Orient. J. Chem.* **2018**, *34*, 548. [CrossRef]
12. Edison, D.J.; Nirmala, W.; Kumar, K.D.A.; Valanarasu, S.; Ganesh, V.; Shkir, M.; AlFaify, S. Structural, optical and nonlinear optical studies of AZO thin film prepared by SILAR method for electro-optic applications. *Phys. B Condens. Matter* **2017**, *523*, 31–38. [CrossRef]
13. Alamro, F.S.; Mostafa, A.M.; Ahmed, H.A.; Toghan, A. Zinc oxide/carbon nanotubes nanocomposite: Synthesis, characterization and catalytic reduction of 4-nitrophenol via laser assistant method. *Surf. Interfaces* **2021**, *26*, 101406. [CrossRef]
14. Alamro, F.S.; Mostafa, A.M.; Al-Ola, K.A.A.; Ahmed, H.A.; Toghan, A. Synthesis of Ag Nanoparticles-Decorated CNTs via Laser Ablation Method for the Enhancement the Photocatalytic Removal of Naphthalene from Water. *Nanomaterials* **2021**, *11*, 2142. [CrossRef]
15. Mostafa, A.M.; Mwafy, E.A.; Awwad, N.S.; Ibrahium, H.A. Au@Ag core/shell nanoparticles prepared by laser-assisted method for optical limiting applications. *J. Mater. Sci. Mater. Electron.* **2021**, *32*, 14728–14739. [CrossRef]
16. Altowyan, A.S.; Mostafa, A.M.; Ahmed, H.A. Effect of liquid media and laser energy on the preparation of Ag nanoparticles and their nanocomposites with Au nanoparticles via laser ablation for optoelectronic applications. *Optik* **2021**, *241*, 167217. [CrossRef]
17. Mostafa, A.M.; Mwafy, E.A.; Awwad, N.S.; Ibrahium, H.A. Catalytic activity of Ag nanoparticles and Au/Ag nanocomposite prepared by pulsed laser ablation technique against 4-nitrophenol for environmental applications. *J. Mater. Sci. Mater. Electron.* **2021**, *32*, 11978–11988. [CrossRef]
18. Mostafa, A.M.; Mwafy, E.A.; Awwad, N.S.; Ibrahium, H.A. Synthesis of multi-walled carbon nanotubes decorated with silver metallic nanoparticles as a catalytic degradable material via pulsed laser ablation in liquid media. *Colloids Surf. A Physicochem. Eng. Asp.* **2021**, *626*, 126992. [CrossRef]
19. Mostafa, A.M.; Mwafy, E.A.; Lotfy, V.F.; Basta, A.H. Optical, electrical and mechanical studies of paper sheets coated by metals (Cu and Ag) via pulsed laser deposition. *J. Mol. Struct.* **2019**, *1198*, 126927. [CrossRef]
20. Mostafa, A.M.; Lotfy, V.F.; Mwafy, E.A.; Basta, A.H. Influence of coating by Cu and Ag nanoparticles via pulsed laser deposition technique on optical, electrical and mechanical properties of cellulose paper. *J. Mol. Struct.* **2020**, *1203*, 127472. [CrossRef]
21. Jelínek, M.; Kocourek, T.; Jurek, K.; Remsa, J.; Mikšovský, J.; Weiserová, M.; Strnad, J.; Luxbacher, T. Antibacterial properties of Ag-doped hydroxyapatite layers prepared by PLD method. *Appl. Phys. A* **2010**, *101*, 615–620. [CrossRef]
22. Pedarnig, J.D.; Heitz, J.; Stehrer, T.; Praher, B.; Viskup, R.; Siraj, K.; Moser, A.; Vlad, A.; Bodea, M.A.; Bäuerle, D.; et al. Characterization of nano-composite oxide ceramics and monitoring of oxide thin film growth by laser-induced breakdown spectroscopy. *Spectrochim. Acta Part B At. Spectrosc.* **2008**, *63*, 1117–1121. [CrossRef]
23. Mostafa, A.M.; Mwafy, E.A. Laser-assisted for preparation Ag/CdO nanocomposite thin film: Structural and optical study. *Opt. Mater.* **2020**, *107*, 110124. [CrossRef]
24. Mostafa, A.M.; Mwafy, E.A. Synthesis of ZnO/CdO thin film for catalytic degradation of 4-nitrophenol. *J. Mol. Struct.* **2020**, *1221*, 128872. [CrossRef]
25. Menazea, A.A.; Mostafa, A.M. Ag doped CuO thin film prepared via pulsed laser deposition for 4-nitrophenol degradation. *J. Environ. Chem. Eng.* **2020**, *8*, 104104. [CrossRef]

26. Srinivasan, N.; Kannan, J.C.; Satheeskumar, S. Examination of antibacterial properties of pure and aluminum doped zinc oxide nanoparticles. *Int. J. Chem. Tech. Res.* **2015**, *7*, 1708–1712.
27. Khatoon, U.T.; Rao, K.V.; Rao, J.R.; Aparna, Y. Synthesis and characterization of silver nanoparticles by chemical reduction method. In Proceedings of the International Conference on Nanoscience, Engineering and Technology (ICONSET 2011), Chennai, India, 28–30 November 2011; pp. 97–99.
28. Alkallas, F.H.; Ahmed, H.A.; Alrebdi, T.A.; Pashameah, R.A.; Alrefaee, S.H.; Alsubhe, E.; Trabelsi, A.B.G.; Mostafa, A.M.; Mwafy, E.A. Removal of Ni(II) Ions by Poly(Vinyl Alcohol)/Al2O3 Nanocomposite Film via Laser Ablation in Liquid. *Membranes* **2022**, *12*, 660. [CrossRef]
29. Alrebdi, T.A.; Ahmed, H.A.; Alkallas, F.H.; Mwafy, E.A.; Trabelsi, A.B.G.; Mostafa, A.M. Structural, linear and nonlinear optical properties of NiO nanoparticles–multi-walled carbon nanotubes nanocomposite for optoelectronic applications. *Radiat. Phys. Chem.* **2022**, *195*, 110088. [CrossRef]
30. Mostafa, A.M.; Mwafy, E.A.; Awwad, N.S.; Ibrahium, H.A. Linear and nonlinear optical studies of Ag/Zn/ZnO nanocomposite thin film prepared by pulsed laser deposition technique. *Radiat. Phys. Chem.* **2021**, *179*, 109233. [CrossRef]
31. Ahmad, M.; Ahmad, I.; Ahmed, E.; Akhtar, M.S.; Khalid, N.R. Facile and inexpensive synthesis of Ag doped ZnO/CNTs composite: Study on the efficient photocatalytic activity and photocatalytic mechanism. *J. Mol. Liq.* **2020**, *311*, 113326. [CrossRef]
32. Pimentel, A.; Araújo, A.; Coelho, B.; Nunes, D.; Oliveira, M.; Mendes, M.; Águas, H.; Martins, R.; Fortunato, E. 3D ZnO/Ag surface-enhanced Raman scattering on disposable and flexible cardboard platforms. *Materials* **2017**, *10*, 1351. [CrossRef]
33. Badawy, M.I.; Mahmoud, F.; Abdel-Khalek, A.A.; Gad-Allah, T.A.; Abdel Samad, A.A. Solar photocatalytic activity of sol–gel prepared Ag-doped ZnO thin films. *Desalinat. Water Treat.* **2014**, *52*, 2601–2608. [CrossRef]
34. Abozied, A.M.; Mostafa, A.M.; Abouelsayed, A.; Hassan, A.F.; Ramadan, A.A.; Al-Ashkar, E.A.; Anis, A. Preparation, characterization, and nonlinear optical properties of graphene oxide thin film doped with low chirality metallic SWCNTs. *J. Mater. Res. Technol.* **2021**, *12*, 1461–1472. [CrossRef]
35. Fageria, P.; Gangopadhyay, S.; Pande, S. Synthesis of ZnO/Au and ZnO/Ag nanoparticles and their photocatalytic application using UV and visible light. *Rsc Adv.* **2014**, *4*, 24962–24972. [CrossRef]
36. Ren, C.; Yang, B.; Wu, M.; Xu, J.; Fu, Z.; Guo, T.; Zhao, Y.; Zhu, C. Synthesis of Ag/ZnO nanorods array with enhanced photocatalytic performance. *J. Hazard. Mater.* **2010**, *182*, 123–129. [CrossRef]
37. Bahedi, K.; Addou, M.; El Jouad, M.; Sofiani, Z.; Alaoui Lamrani, M.; El Habbani, T.; Fellahi, N.; Bayoud, S.; Dghoughi, L.; Sahraoui, B.; et al. Diagnostic study of the roughness surface effect of zirconium on the third-order nonlinear-optical properties of thin films based on zinc oxide nanomaterials. *Appl. Surf. Sci.* **2009**, *255*, 4693–4695. [CrossRef]
38. Zhao, S.; Zhou, Y.; Zhao, K.; Liu, Z.; Han, P.; Wang, S.; Xiang, W.; Chen, Z.; Lü, H.; Cheng, B. Violet luminescence emitted from Ag-nanocluster doped ZnO thin films grown on fused quartz substrates by pulsed laser deposition. *Phys. B Condens. Matter* **2006**, *373*, 154–156. [CrossRef]
39. Khattab, T.A.; Gabr, A.M.; Mostafa, A.M.; Hamouda, T. Luminescent plant root: A step toward electricity-free natural lighting plants. *J. Mol. Struct.* **2019**, *1176*, 249–253. [CrossRef]
40. Alkallas, F.H.; Toghan, A.; Ahmed, H.A.; Alrefaee, S.H.; Pashameah, R.A.; Alrebdi, T.A.; Mwafy, E.A.; Mostafa, A.M. Catalytic performance of NiO nanoparticles decorated carbon nanotubes via one-pot laser ablation method against methyl orange dye. *J. Mater. Res. Technol.* **2022**, *18*, 3336–3346. [CrossRef]
41. Altowyan, A.S.; Toghan, A.; Ahmed, H.A.; Pashameah, R.A.; Mwafy, E.A.; Alrefaee, S.H.; Mostafa, A.M. Removal of methylene blue dye from aqueous solution using carbon nanotubes decorated by nickel oxide nanoparticles via pulsed laser ablation method. *Radiat. Phys. Chem.* **2022**, *198*, 110268. [CrossRef]
42. Mwafy, E.A.; Mostafa, A.M.; Awwad, N.S.; Ibrahium, H.A. Catalytic activity of multi-walled carbon nanotubes decorated with tungsten trioxides nanoparticles against 4-nitrophenol. *J. Phys. Chem. Solids* **2021**, *158*, 110258. [CrossRef]

membranes

MDPI

Article

Removal of Ni(II) Ions by Poly(Vinyl Alcohol)/Al_2O_3 Nanocomposite Film via Laser Ablation in Liquid

Fatemah H. Alkallas [1], Hoda A. Ahmed [2,3], Tahani A. Alrebdi [1], Rami Adel Pashameah [4], Salhah H. Alrefaee [3], Emaan Alsubhe [5], Amira Ben Gouider Trabelsi [1], Ayman M. Mostafa [6,7,*] and Eman A. Mwafy [7,8]

1 Department of Physics, College of Science, Princess Nourah bint Abdulrahman University, P.O. Box 84428, Riyadh 11671, Saudi Arabia; fhalkallas@pnu.edu.sa (F.H.A.); taalrebdi@pnu.edu.sa (T.A.A.); aatrabelsi@pnu.edu.sa (A.B.G.T.)

2 Department of Chemistry, Faculty of Science, Cairo University, Cairo 12613, Egypt; ahoda@sci.cu.edu.eg

3 Chemistry Department, College of Sciences, Taibah University, Yanbu 30799, Saudi Arabia; srfaay@taibahu.edu.sa

4 Department of Chemistry, Faculty of Applied Science, Umm Al-Qura University, Makkah 24230, Saudi Arabia; rapasha@uqu.edu.sa

5 Physics Department, Faculty of Science, Taibah University, Yanbu 30799, Saudi Arabia; esobhe@taibahu.edu.sa

6 Spectroscopy Department, Physics Division Institute, National Research Centre, 33 El Bohouth St. (Former El Tahrir st.), Dokki, Giza 12622, Egypt

7 Laser Technology Unit, Center of Excellent for Advanced Science, National Research Centre, 33 El Bohouth st. (Former El Tahrir St.), Dokki, Giza 12622, Egypt; emanmwafynrc@gmail.com

8 Physical Chemistry Department, Advanced Materials Technology and Mineral Resources Research Institute, National Research Centre, Giza 12622, Egypt

* Correspondence: aymanmdarwish@gmail.com

Abstract: Al_2O_3-poly(vinyl alcohol) nanocomposite (Al_2O_3-PVA nanocomposite) was generated in a single step using an *eco*-friendly method based on the pulsed laser ablation approach immersed in PVA solution to be applicable for the removal of Ni(II) from aqueous solution, followed by making a physicochemical characterization by SEM, XRD, FT-IR, and EDX. After that, the effect of adsorption parameters, such as pH, contact time, initial concentration of Ni(II), and medium temperature, were investigated for removal Ni(II) ions. The results showed that the adsorption was increased when pH was 5.3, and the process was initially relatively quick, with maximum adsorption detected within 90 min of contact time with the endothermic sorption process. Moreover, the pseudo-second-order rate kinetics ($k_2 = 9.9 \times 10^{-4}$ g mg^{-1} min^{-1}) exhibited greater agreement than that of the pseudo-first-order. For that, the Ni(II) was effectively collected by Al_2O_3-PVA nanocomposite prepared by an *eco*-friendly and simple method for the production of clean water to protect public health.

Keywords: PLAL; Al_2O_3; NPs; Nd:YAG; laser ablation; nanocomposite

1. Introduction

Nano adsorbent materials have sparked tremendous attention in recent years because of their large specific surface area, regular pore structure, and highly controlled surface characteristics. Almost the majority of the adsorbents that are created today to overcome the hazard of heavy metal ions and dyes depend on the engagement of the hazardous chemical compounds with the presence of functional groups on the adsorbents' surfaces [1–5]. As a result, the matrix's vast surface area and many adsorption sites are the most critical elements influencing the adsorbent's adsorption capacity. From hazardous chemical compounds, nickel waste is extremely dangerous to people, ecosystems, and animals. It has been found in a wide range of industrial wastes, including nickel–cadmium batteries, organic compounds, and insecticides [6–9]. The abundance of nickel ions in the water supply has been linked to a variety of major health issues, including carcinogenic agents. As a result, removing nickel from wastewater and industrial pollutants is critical [10–13]. Several techniques have been employed to remove heavy metals from wastewater, including chemical precipitation, ion-exchange sorption, and solvent extraction. One of these methods, the adsorption method has been frequently employed because of its low cost and great efficacy [14–19].

Aluminum oxide nanostructured material (Al_2O_3) has been studied as a promising adsorbent for heavy metal ion elimination because of its large surface area. Moreover, in the water treatment field, it exhibits better catalytic activity in a variety of organic processes [20,21]. However, in the field of water treatment, powdered material is not suited for recycling. Furthermore, the separation of nanostructured materials from aqueous solutions is quite challenging, especially when the particle size is on the nano-scale. Researchers are currently concentrating on methods of employing nano-powder metal oxides to remove heavy metal ions [22–24]. One method is to incorporate nanoparticles into an organic substance, which could be represented by an optimum solution such as a polymer matrix structure. Adsorbent features such as surface area and functional groups on the adsorbent surface should be addressed for the adsorbent to eliminate heavy metal ions. The important qualities of the adsorbent, such as high mechanical and thermal stability and high sorption capacity, should be carefully addressed. For these reasons, selecting appropriate materials for heavy metal ion sorption from an aqueous medium is critical. Adsorbents such as zeolites, polymers, metal oxides, and polymer/metal oxides have been used as materials for the sorption of heavy metal ions. From these promising absorbents, poly (vinyl alcohol) PVA is identified to be associated with various organic and inorganic compounds able to supply hydroxyl groups on its surface for the elimination of heavy metal ions. Because of its unique features such as high porosity, large surface area, and tiny pore size, it has been widely employed for heavy metal ion adsorption. Furthermore, it provides a greater possibility for nanoparticle loading without affecting mechanical stability [25–27].

Metal oxide nanoparticles placed inside a polymer matrix have recently piqued the interest of researchers due to the unique features of hybrid nanocomposites. Many factors influence the properties of the created nanocomposites, including the synthesis method, its morphological structure, the characteristics of the metal oxide nanoparticles that are implemented in PVA, and the physicochemical properties of both metal oxide and polymer molecules in the interaction [28–30]. One of these techniques, that helped us to vary the characteristics of embedding metal oxide in a simple way, is the pulsed laser ablation in liquid media approach. It is a simple, quick, economical, green, dependable method for producing a wide range of nanostructured materials and a commonly used technology for producing nanocomposite adsorbent materials that has received a lot of attention recently. This method allows us to produce nanostructured materials in high purity without any minor products as it is based on using just required precursor pure metals not in their salts at the beginning of the preparation, and it did not require surfactants to preserve the size in nanoscale form. Furthermore, the normal behavior of the preparation of nanostructured materials via any physical or chemical approach methods is associated with the generation of hazard vapor fumes consisting of very finer powders from the prepared nanostructured form, which have a very dangerous side effect on the environment and human health as they are capable of passing through human skin, while in this method, all the preparation steps were carried out inside a liquid medium, which prevents the transfer of any vapor fumes to the environment or human bodies. Therefore, this method could be considered an *eco*-friendly and green method. Furthermore, this method is eligible to form nanocomposite structures by ablating many metal targets at the same time or changing the liquid media with a matrix structure, such as PVA or CNTs. Therefore, this method fits the need for versatility, which affects the shape and size of the produced nanostructured materials in comparison to other methods. However, this mass production of this method could represent the main issue faced by the enhancement and development of this method [31–35]. The achievement of the PLAL method in the preparation of Al_2O_3 NPs could be summarized as follows: In 2010, S. Khan et al. studied the effect of the ablation time of the continuous wave (CW) of fiber laser to obtain Al_2O_3 NPs with different particle sizes [36]. In 2012, V. Piriyawong et al. prepared Al_2O_3 NPs by ablating Al pellet immersed in deionized water. It was shown that their particle size could be changed by varying the ablation energy [37]. In 2020, A. Riahi et al. showed the prepared Al_2O_3 NPs by laser ablation (7 ns, 1064 nm, and 120 mJ) of Al target immersed in deionized water with have high thermal conductivity [38]. Moreover, M. Eskandari et al. prepared Au@Al_2O_3 core-shell in just one step by ablating bulk aluminum deposited with a thin layer from the Au layer immersed in ethanol solution by CW fiber laser (1064 and 40 W). This preparation method could be represented the development of the PLAL method for generating core/shell nanocomposite structures [39]. Furthermore, A. Mostafa et al. showed the capability for preparing different metal oxides embedded in PVA structure in just a one-pot method by ablating metal plates (Al, Cd, and Cu) in PVA solution [40]. Moreover, A. Mostafa et al. showed another facility laser parameter to enhance the PLAL method, by making laser irradiation after the laser ablation process to decrease the particle size of the prepared Al_2O_3 NPs [41]. Furthermore, the previous contributions of Al_2O_3-PVA nanocomposite for the removal of heavy metals or degraded organic pollutants can be summarized in Table 1.

Table 1. Recent survey of nanocomposite based on Al_2O_3-PVA for removal heavy metals or degradation of organic pollutants.

Composite	Organic Pollutants/Heavy Metals	Efficacy
Al_2O_3-PVA [42]	phosphate	95%
PVA–ZnO–Al2O_3 [43]	MB	100%
Polythiophene/PVA/Al_2O_3 [44]	Pb(II), Zn(II), Cd(II)	97.3% 89.4% 95.8%
Polyaniline/PVA/Al_2O_3 [44]	Pb(II), Zn(II), Cd(II)	89.78% 84.9% 79.2%

In this work, an Al_2O_3-PVA nanocomposite was effectively produced in just one step using the *eco*-friendly promising tools of the PLAL method to be suitable for nickel ion adsorption from aqueous solutions. The properties of the synthesized hybrid structure were carried out by different techniques. After that, the influence of several factors on the sorption process, such as concentration, contact duration, beginning concentration, and temperature was examined for Ni(II).

2. Materials and Experimental Work

2.1. Materials

Ultra-pure water was used to make all of the reagents. Poly vinyl alcohol solution (PVA) was purchased from La-113 laboratory, Rasayan, Egypt. Aluminum sheet (Al) with dimension 2×2 cm^2 and thickness of 0.1 mm was purchased from the BDH Chemical Ltd. pool, England. Nickel nitrate ($NiNO_3$) was purchased from LOBA Chemie, Laboratory Reagents and fine Chemicals, India, which was used to make a stock solution of Ni(II) ions. Sodium hydroxide (NaOH) and hydrochloric acid (HCl) were purchased from El Nasr Pharmaceutical Co., Giza, Egypt.

2.2. Preparation of PVA Solution

A 10 wt.% PVA was produced by dissolving 10 g of PVA in 100 mL of ultra-pure water and maintaining it at 80 °C for 300 min with continued magnetic stirring.

2.3. Preparation of Al_2O_3-PVA Composite

The Al_2O_3 nanostructured material was generated by the pulsed laser ablation of cleaned metal targets of Al sheet put in a glass vessel filled with 5 mL of the prepared PVA solution to form an Al_2O_3 nanostructured material embedding PVA. The laser ablation process produced by the laser beam of the first harmonic generation of nanosecond Nd:YAG laser, which produces a 10 Hz repetition rate and 150 mJ of laser energy. The ablation process was carried out by focusing the laser beam on the target's surface with a planco-convex lens (70 mm) to produce a small spot that had a 20 μm diameter (Figure 1). The ablation procedure was carried out under mechanical stirring for 30 min to create distinct nanocomposite structures. Then, the prepared nanocomposite solution was separately cast in a glass Petri-dish to obtain the required films, followed by drying in glass Petri dishes with a thickness of about 2 mm in the form of free-standing and a diameter of 25 mm. The amount of immobilized Al_2O_3 NPs in PVA could be estimated in two ways: the first method was related to measuring the weight of the Al-sheet before and after the ablation process, followed by subtraction of the two values ($\text{weight}_{\text{Al sheet}} - \text{weight}_{\text{ablated Al sheet}}$) to estimate the amount of immobilized Al_2O_3 NPs in PVA. The second method was related to measuring the weight of PVA film before and after being embedded with Al_2O_3 NPs, followed by subtraction of the two values ($\text{weight}_{\text{embedded PVA film}} - \text{weight}_{\text{pure PVA film}}$) to estimate the amount of immobilized Al_2O_3 NPs in PVA. After using both methods and repeating each method three times, it was detected that the amount of embedded Al_2O_3 in the PVA film is 9.2 μg ± 0.051.

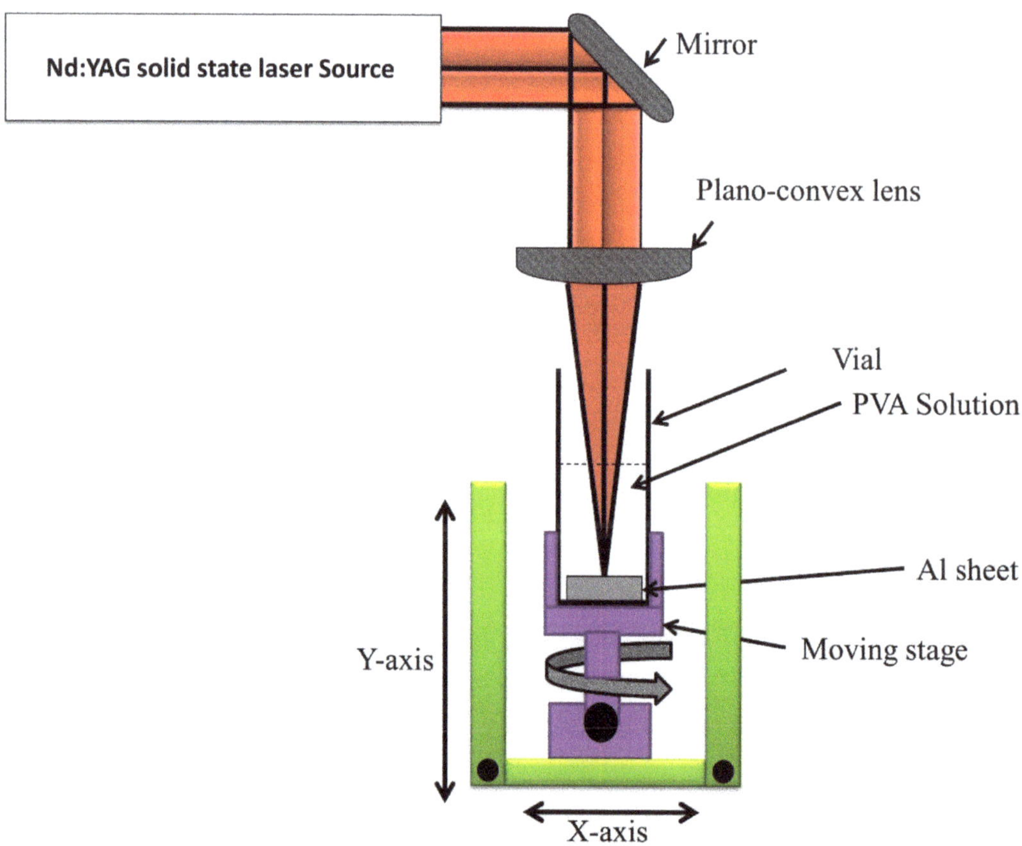

Figure 1. Schematic diagram of Al_2O_3-PVA nanocomposite by PLAL.

2.4. Investigation Techniques

X-ray diffractometer (Shimadzu XRD 7000, Kyoto, Japan), UV–VIS-NIR spectrophotometer (JASCO 570 UV–VIS-NIR, Kyoto, Japan), atomic absorption spectroscopy (Shimadzu AA-6200; Shimadzu Co., Ltd., Kyoto, Japan), FT-IR (JASCO FT-IR 6100 spectrometer, Japan), and a scanning electron microscope associated with EDX analyzer detector (PHILIPS/FEI QUANTA 250, Prague, Czech Republic).

2.5. Adsorption Study

The batch sorption of Ni(II) ions onto a PVA-Al_2O_3 nanocomposite adsorbent in aqueous solutions at ambient temperature was examined. For that, a conical flask containing 100 mL of Ni^{2+} metal ion solution was filled with ultra-pure water and stirred to achieve an initial Ni solution concentration ranging from 10 to 100 mg/L before being put in a thermostatic shaker for absorption (200 rpm). The remaining Ni^{2+} ion concentration was then measured using atomic absorption spectroscopy. The following formulae were used to investigate the effect of various parameters on the adsorption capacity of Ni^{2+} [45,46].

$$\text{Removal extent, percentage} = (C_o - C_a)/C_o \times 100 \quad (1)$$

where C_o and C_a are the concentrations of adsorbent material at the beginning of the adsorption process and at the equilibrium of the adsorption process, respectively.

3. Result and Discussion

3.1. Investigation of the Prepared Al_2O_3-PVA Nanocomposite

Figure 2 shows the FT-IR spectra of PVA, Al_2O_3 nanoparticles, and Al_2O_3-PVA nanocomposite to identify their vibrational absorption motion, which is produced by FT-IR spectra. The FT-IR spectra of pure PVA exhibited a strong and wide stretching band at 3334 cm^{-1} and a bending vibration at 1565 cm^{-1}, 1428 cm^{-1}, and 1325 cm^{-1} that can be attributed to the hydroxyl group because of intramolecular and intermolecular hydrogen bonding. Furthermore, stretching vibrational absorption bands at 2945 cm^{-1} and 2908 cm^{-1} appeared as a result of functional groups of the CH and $-CH_2$ groups, respectively. Furthermore, the appearance of stretching vibration around 1100 cm^{-1} was related to the C-O functional group. These characteristics could be represented by the main functional groups of the PVA structure. In addition to the appearance of a vibrational stretching peak around 1739 cm^{-1} related to the C=O functional group, but with small intensity as it was related to the presence of leftover acetate groups after the hydrolysis of polyvinyl acetate to produce PVA [30,47,48]. In the case of the laser ablation of the Al sheet immersed in the ultra-pure water, the FT-IR spectra of Al_2O_3 nanoparticles exhibited two distinctive vibrational peaks of Al_2O_3, at 558 cm^{-1} and 632 cm^{-1}, which are exhibited in the fingerprint region. Furthermore, the appearance of vibrational peaks characterized by a broad band stretching motion around 3342 cm^{-1} was related to the residual hydroxyl functional group (-OH) that was produced by atmospheric moisture. Moreover, there are no other peaks related to the residual organic compounds. In the case of Al_2O_3-PVA nanocomposite, the primary distinctive vibrational peaks of the PVA structure are still intact, but their strength and position have changed. The intensity of these hydroxyl group bands is clearly reduced, which is attributable to the interaction of the PVA structure with Al ions. Moreover, below 1000 cm^{-1}, there are two distinct peaks of Al_2O_3, at 558 cm^{-1} and 632 cm^{-1}, which are exhibited in the fingerprint area of the spectra as evidence for the incorporation of Al_2O_3 molecules in the polymer matrix. Moreover, it was clear that the strong electronegativity of the atoms formed from Al_2O_3 and injected into the polymeric matrix of PVA structure, which has a significant impact on the spectrum of adjacent group frequencies, may be responsible for the conformational changes in the width and intensity of the vibrational bands of PVA structure [20,41,49]. These significant fluctuations in the absorbance and frequencies of almost bands suggest that interactions between Al_2O_3 NPs and the polymeric structure of PVA, such as hydrogen bonds or van der Waals interactions, are forming between the (OH^-)/(COO^-) groups of PVA and the Al_2O_3 NPs. This was accomplished by reducing the crystallinity degree of nanocomposites and forming charge transfer complexes in which the Al_2O_3 NPs function as an electron acceptor and the PVA matrix works as an electron donor. These interactions alter the dynamics of the nanocomposite chain structure.

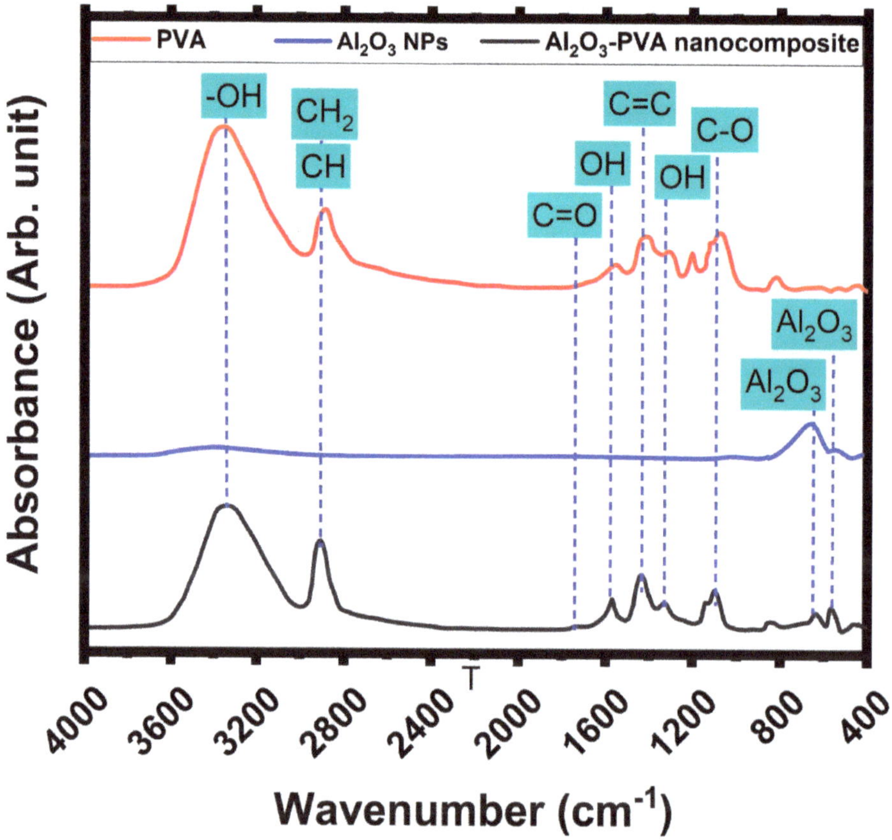

Figure 2. FT-IR spectra of PVA, Al_2O_3 nanoparticles, and Al_2O_3/PVA nanocomposite.

FESEM was used to investigate the morphological changes in PVA caused by embedding with Al_2O_3 molecules as mentioned in Figure 3. The micrographs exhibit SEM of Al_2O_3 nanoparticles, PVA, and the hybrid with PVA-Al_2O_3 nanocomposite. The surface of the produced structure from the laser ablation of the Al sheet in the ultra-pure water showed a semi-spherical shape in the nanoscale form, while the SEM image of the PVA structure showed a uniform surface shape that reveals dispersed smooth microspores instead of uniform ones. Moreover, the surface of the polymer was significantly distorted during metal oxide contact, and the metal oxide particles were uniformly dispersed over the polymer surface compared to the pure PVA structure. As a result, the change in the surface of the PVA-Al_2O_3 nanocomposite reveals that the PVA surface was highly deformed, and the aluminum oxides were homogeneously distribution throughout the PVA surface.

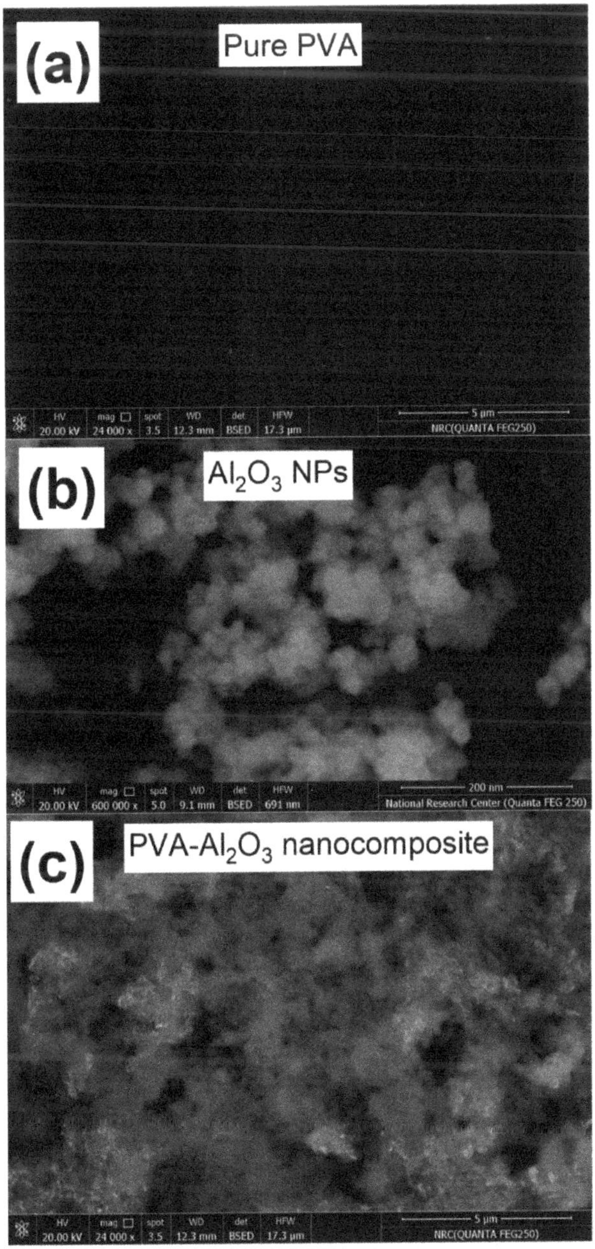

Figure 3. SEM image of (**a**) PVA, (**b**) Al_2O_3 nanoparticles, and (**c**) Al_2O_3/PVA nanocomposite.

Furthermore, the elements on the surface of the PVA structure before and after embedding with Al_2O_3 NPs were examined using an EDX element analyzer. Figure 4 depicts the EDX spectra of a PVA-Al_2O_3 nanocomposite before and after Al_2O_3 NPs adsorption. This result verified the presence of a pure PVA structure as it was only comprised of the C and O elements, which represented the main constituted elements of the PVA structure. Moreover, the elemental analysis confirmed the formation of pure Al_2O_3 nanoparticles from the laser ablation process in the ultra-pure water as it was only comprised of the Al and O elements, which represented the main constituted elements of the Al_2O_3 structure. However, in the case of embedding with Al_2O_3, a new element (Al) was injected with C and O. So, the PVA structure was embedded with the Al element. Moreover, the phase of Al could be detected to be metallic or oxide form based on looking at the ratio of O/C in PVA structure before and after ablation of Al, which was 0.797 and 1.42, respectively. So, the amount of O was increased owing to the occurrence of Al in the oxide form (Al_2O_3).

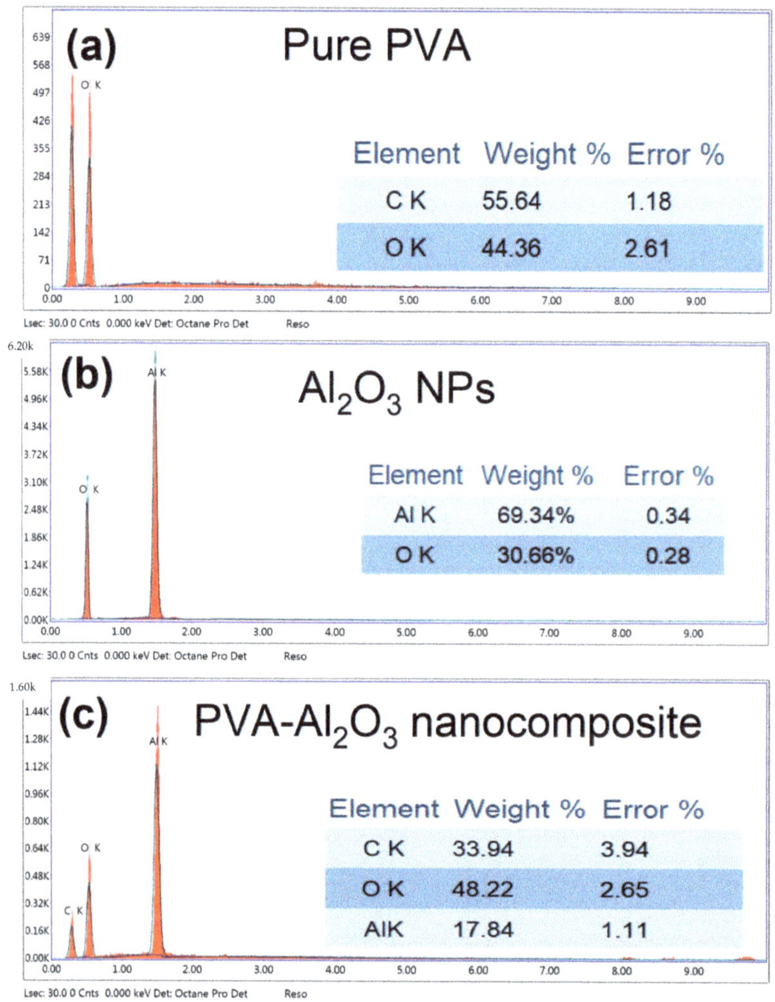

Figure 4. EDX elemental analysis of (**a**) PVA and (**b**) EDX elemental analysis of PVA before and (**c**) after embedding with Al_2O_3.

The XRD measurements were used to examine the crystallinity of the PVA structure with and without incorporated metal produced from the laser ablation of the Al sheet. Nanostructured materials were created by the PLAL method of an Al target immersed in different types of liquid media, and the resultant diffractogram spectra are given in Figure 5. In the case of pure PVA, there are two separate diffraction peaks at diffraction angles $2\theta = 19.5°$ and $40.5°$, which correspond to the indices planes (1 0 1) and (1 1 1), respectively [50]. Furthermore, no other peaks were observed related to any foreign impurities in the PVA structure.

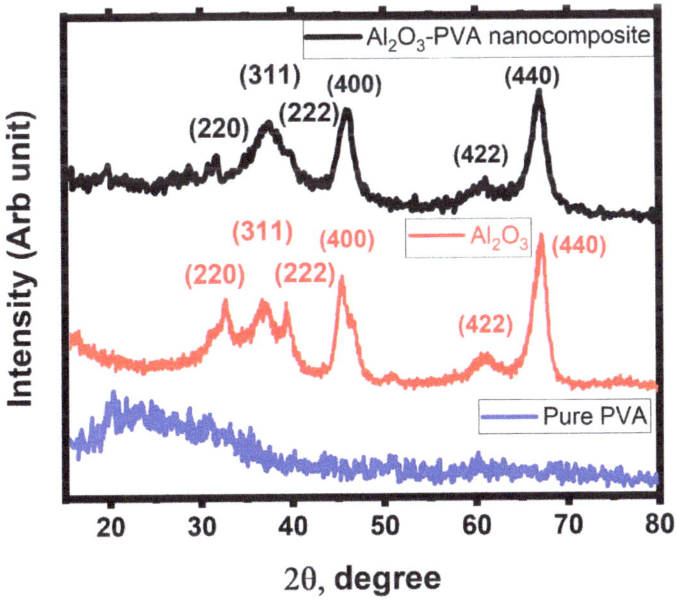

Figure 5. XRD diffractogram of PVA, Al_2O_3 nanoparticles, and Al_2O_3/PVA nanocomposite.

In the case of the laser ablation of the Al sheet immersed in the ultra-pure water, the crystalline peaks appeared at 31.62°, 37.18°, 39.79°, 46.17°, 60.84°, and 67.13°, which correspond to (2 2 0), (3 1 1), (2 2 2), (4 0 0), (4 2 2), and (4 4 0), respectively, based on JCPD XRD card number 79-1558 of the cubic structure of Al_2O_3. Moreover, no other typical peaks were seen in the diffractogram, which may be produced from the other contaminants during the preparation steps. This observation confirmed the high purity of the produced nanostructure materials. Moreover, the main crystalline peak used to deetect the crystallite size of Al_2O_3 nanostructure using the Debye–scherrer equation [51].

$$D = \frac{0.9\lambda}{\beta \cos\theta} \tag{2}$$

where D, λ, β, and θ are the crystallite size of Al_2O_3 nanostructure, used X-ray wavelength, FWHM of the diffraction peak at 2θ = 67.13°, and the Bragg angle (33.57°), respectively, its crystallite is 19.8 nm. In the case of laser ablation of Al sheet immersed in the PVA solution, the semi-crystalline structure was still present in the case of embedding with the generated nanostructured materials formed from the PLAL process of Al-target immersed in PVA solution, but the intensity was reduced by a large amount under the effect of the intermolecular hydrogen band produced between the PVA structure and nanostructured materials. Furthermore, the new crystalline peaks appeared at 31.62°, 37.18°, 39.79°, 46.17°, 60.84°, and 67.13°, which correspond to (2 2 0), (3 1 1), (2 2 2), (4 0 0), (4 2 2), and (4 4 0), respectively, based on JCPD XRD card number 79-1558 of the cubic structure of Al_2O_3 [52]. It was clear that the PVA diffraction peaks remained after being incorporated with Al_2O_3 NPs in the creation of a nanocomposite structure. It was demonstrated that the Al_2O_3 NPs incorporation had no effect on the distribution of the PVA's crystallinity, while its area under the curve has changed to a lower value, indicating that the crystallinity of the PVA structure has reduced. This reduction could be related to the presence of intermolecular hydrogen bonding that occurred between its chain structure and Al_2O_3 NPs, leading to reducing the intermolecular interaction between the chains of PVA. Therefore, the electrostatic interaction of Al_2O_3 NPs with the PVA chain leads to disturbing the crystalline phase of the PVA structure. So, this observation could be represented as a confirmation of successful incorporation.

3.2. *Adsorption Process*

3.2.1. Effect of pH

The pH of the solution is a critical variable in the adsorption process because it influences metal sorption via protonation and deprotonation of the adsorbent's functional groups, the Ni speciation, adsorbent surface charge, and adsorbent ionization. As a result, the influence of pH on nickel sorption is explored in the pH range 2–7 for an initial concentration of 100 mg L^{-1} from Ni(II), an adsorbent dose of 0.5 g/L, and a temperature of study at 303 K as depicted in Figure 6a. This graph clearly shows that the adsorption capacity rises with a rising pH value, and the precipitation occurs in nickel solutions when the pH exceeds 7. So, the trials are not undertaken above a pH of 7. Furthermore, around pH 5.3, nickel adsorption capacity reaches a maximum of 62 mg g^{-1} and thereafter decreases with increasing pH levels [53–56]. This was related to:

1. At pH values higher than 6, the excess of alkaline –OH group has a greater tendency to combine with Ni^{2+} and form $Ni(OH)_2$ participated, causing the adsorption to be diminished.
2. At pH values in the range 5–6, the surface charge of the hybrid nanocomposite turned to a negative charge due to the medium having a low acidic concentration, increasing the coordination between positively charged metal ions (Ni^{2+}) and Al_2O_3-PVA nanocomposite by electrostatic attraction, leading to reach the maximum adsorption capacity. The results were consistent with Ni^{2+} adsorption on Al_2O_3-PVA nanocomposite as mentioned in the previous work [57].
3. At pH values lower than 5, the decrease in nickel adsorption capability at lower pH levels is related to protonation of the Al_2O_3-PVA nanocomposite by the acidic medium and the water molecules converted from H_2O to H_3O^+, leading to a decrease in the number of charge carriers in the hybrid membrane for metal adsorption. In addition, competition for adsorption sites on the PVA structure created between H^+ and Al^{2+} make an electrostatic repulsion between them [58–61].

3.2.2. Effect of Initial Pollutant Concentration

The influence of initial Ni^{2+} concentrations on nanocomposite adsorption was examined from 0 mgL^{-1} to 500 mgL^{-1} of beginning Ni(II) concentration at a temperature of study at 303 K, pH 5.3, and an adsorbent dose of 0.5 g/L, and the findings are presented in Figure 6b. It was shown that the initial Ni^{2+} concentration in aqueous solution increased till it reached saturation and then decreased. So, the precision description of this process can be described as the percentage amount of Ni(II) removal greatly reduced as the beginning Ni(II) ion concentration content rises. This observation could be discussed when the initial Ni(II) ion concentration increased from 0 to 200 mg/L, the mass transfer of Ni(II) ions between the aqueous solution and the Al_2O_3 nanoparticles were supported, leading to improved interaction between Ni(II) ions and the Al_2O_3 nanoparticles and increasing the adsorption uptake of Ni(II). In other words, when the initial Ni(II) ion concentration was further increased from 200 to 500 mg/L, the adsorption capacity hit a plateau due to the strong binding sites on the Al_2O_3/PVA nanocomposites have been filled with the initial amounts of Ni(II) ions, which appeared from the adsorption isotherm analysis as the Langmuir and Freundlich models, which were compatible with the other scholars who had observed similar findings. So, from this study, the optimum beginning Ni(II) ion concentration for a good adsorption process was chosen before the plateau shape, at 100 mg/L as this value represented the highest rate of adsorption capacity and after this value, the rate of adsorption capacity started to be decreased [62–64].

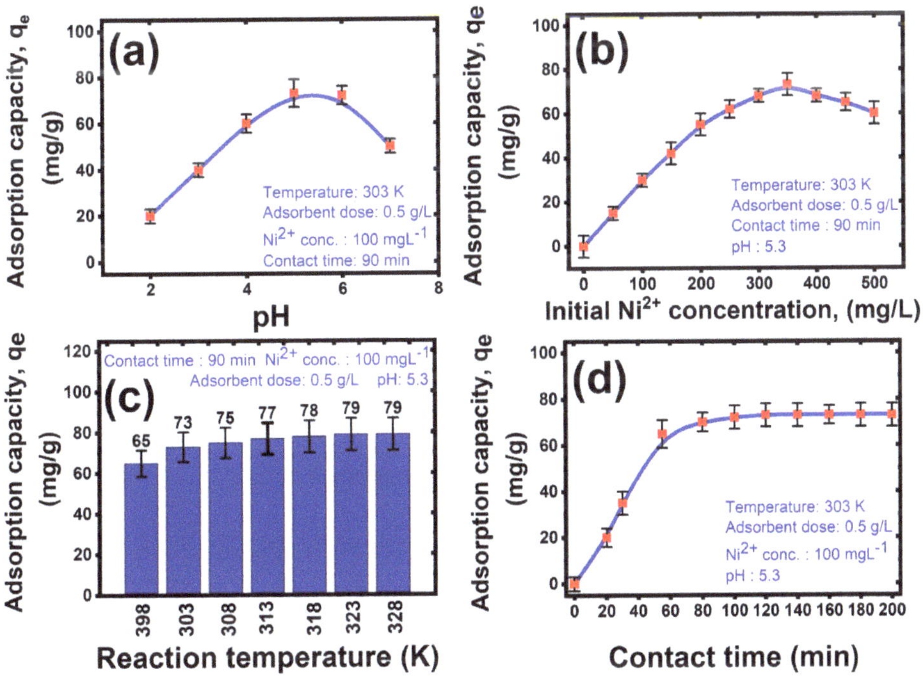

Figure 6. Effect of (**a**) pH, (**b**) beginning Ni^{2+} concentration, (**c**) reaction temperature, and (**d**) contact time on Ni(II) removal by Al_2O_3/PVA nanocomposite.

3.2.3. Effect of Temperature

Figure 6c depicts the influence of reaction temperature on the removal of Ni^{2+} ions for 100 mg L^{-1} beginning Ni(II) ion concentration, an adsorbent dose of 0.5 g/L, and pH of 5.3 for nickel solutions at varying temperatures from 298 K to 328 K. As the temperature rose from 298 to 328 K, the adsorption capacity of nickel ions grew from 65.8 mg g^{-1} to 87.4 mg g^{-1}. These findings revealed that nickel ion sorption onto the Al_2O_3/PVA nanocomposite adsorbent was endothermic.

3.2.4. Effect of Contact Time

Figure 6d depicts the influence of contact time on the removal of Ni^{2+} ions for starting metal concentrations of 0 and 500 mgL^{-1}, an adsorbent dose of 0.1 g/L, an adsorbent dose of 0.5 g/L, and pH of 5.3 for nickel solutions at 303 K temperature. At varied starting metal concentrations of 0 and 100 ppm, the adsorption efficiency for nickel ions by the Al_2O_3/PVA nanocomposite rose dramatically and achieved equilibrium after 90 min. As a result, a time of 50 min was chosen as an equilibrium time for future tests. Furthermore, as demonstrated, the sorption capacity of Al_2O_3/PVA nanocomposite increases with time due to the increased diffusivity of nickel ions toward Al_2O_3/PVA nanocomposite [65].

3.3. Mechanism of Adsorption Process

The adsorption of metal ions of Ni^{2+} by the nanocomposite could be controlled in three ways. The first predicted mechanism was the nanocomposite adsorbent's quick transfer to the external surface absorption of metal ions. The second predicted mechanism was the diffusion of metal ions via the pores of the nanocomposite. The third predicted mechanism might be linked to chemical adsorption between nanocomposite functional groups, such as hydroxyl groups, and metal ions. The experimental results were analyzed using pseudo-first-order or pseudo-second-order kinetic models to study the kinetic parameters of the adsorption of metal ions via the Al_2O_3/PVA nanocomposite [30,66–68].

$$q_t = q_e\left(1 - e^{-k_1 t}\right) \text{ Pseudo-first-order} \tag{3}$$

$$q_t = \left(k_2 q_e{}^2 t\right)/(1 + q_e k_2 t) \text{ Pseudo-second-order} \tag{4}$$

where q_t is the amount of adsorbed metal after t time, q_e is the amount of adsorbed metal after equilibrium time, k_1 is the constant of pseudo-first-order, and k_2 is the constant of pseudo-second-order. From Figure 7, the intercept, slope, and the correlation coefficient of Pseudo-first-order are 4.95347 ± 0.03878, −0.3606 ± 0.00983, and 0.98969, respectively, while the intercept, slope, and the correlation coefficient of Pseudo-second-order are 0.15013 ± 0.00719, 0.01219 ± 8.87805 × 10^{-5}, and 0.99926, respectively. So, the value of k_1 and k_2 are 0.3606 and 9.9 × 10^{-4} gmg^{-1} min^{-1}, respectively. Therefore, in the case of the pseudo second-order, the predicted equilibrium adsorption capacity was more compatible with its experimental value.

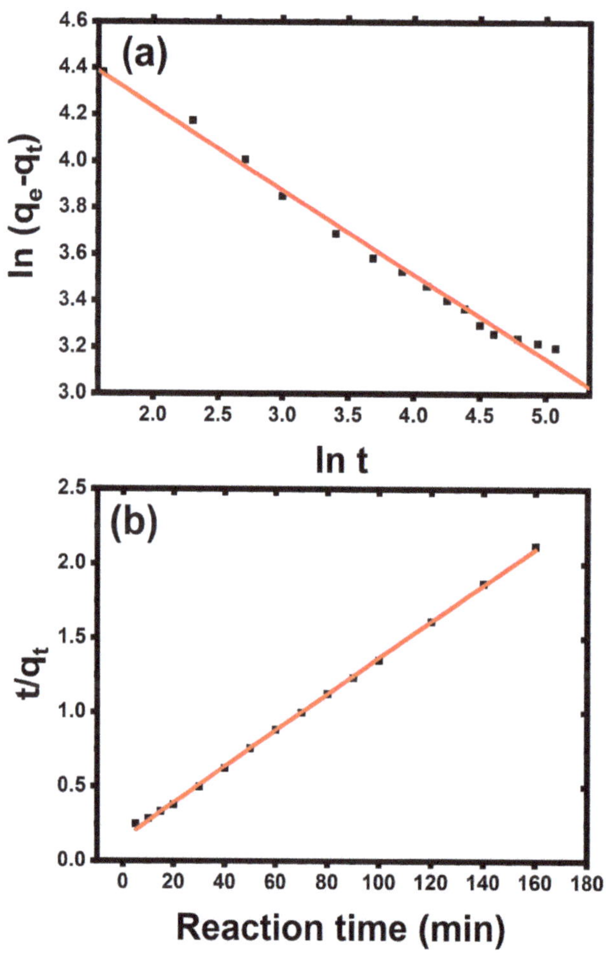

Figure 7. Adsorption kinetics of Ni(II) ions by Al_2O_3/PVA nanocomposite (**a**) Pseudo-first-order and (**b**) Pseudo-second-order.

4. Conclusions

A novel hybrid PVA/Al_2O_3 nanocomposite was created in this study using a simple green potential approach. This approach used a pulsed laser ablation method to create a hybrid nanocomposite from an Al sheet immersed in PVA solution. The successful embedding of PVA structure with Al_2O_3 nanoparticles was demonstrated by FTIR, XRD, and SEM-EDAX investigation of PVA before and after contact with Al_2O_3 nanoparticles. It can be utilized as a cost-effective alternative to activated carbons, membrane filtration, and ion-exchange adsorbents. The adsorbent effectively removes Ni(II) from aqueous solutions, with even greater removal achieved at low concentrations of about 100 mg/L. The adsorption was shown to be pH-dependent, with pH 5.3 yielding the most clearance. The pseudo-second-order rate kinetics was used to manage the removal process. The adsorption findings revealed that hybrid PVA/Al_2O_3 nanocomposite may be utilized to efficiently remove Ni(II) from aqueous solutions.

Author Contributions: Conceptualization, A.M.M., F.H.A. and E.A.M.; methodology, S.H.A., E.A., E.A.M. and F.H.A.; software, H.A.A., T.A.A., F.H.A. and A.M.M.; formal analysis, A.M.M. and H.A.A.; investigation, A.M.M., R.A.P., S.H.A., E.A. and E.A.M.; resources, T.A.A., A.B.G.T. and F.H.A.; data curation, H.A.A.; writing—original draft preparation, A.M.M., E.A.M. and H.A.A.; writing—review and editing, F.H.A. and R.A.P.; supervision, A.M.M.; project administration, A.M.M. and F.H.A.; funding acquisition, F.H.A. All authors have read and agreed to the published version of the manuscript.

Funding: This research was funded by Princess Nourah bint Abdulrahman University Researchers Supporting Project number (PNURSP2022R38), Princess Nourah bint Abdulrahman University, Riyadh, Saudi Arabia.

Institutional Review Board Statement: Not Applicable.

Informed Consent Statement: Not Applicable.

Data Availability Statement: Not Applicable.

Acknowledgments: The authors express their gratitude to Princess Nourah bint Abdulrahman University Researchers Supporting Project number (PNURSP2022R38), Princess Nourah bint Abdulrahman University, Riyadh, Saudi Arabia. Also, the authors would like to thank the Deanship of Scientific Research at Umm Al-Qura University for supporting this work by Grant Code: (22UQU4320141DSR09).

Conflicts of Interest: The authors declare no conflict of interest.

References

1. Altowyan, A.S.; Toghan, A.; Ahmed, H.A.; Pashameah, R.A.; Mwafy, E.A.; Alrefaee, S.H.; Mostafa, A.M. Removal of methylene blue dye from aqueous solution using carbon nanotubes decorated by nickel oxide nanoparticles via pulsed laser ablation method. *Radiat. Phys. Chem.* **2022**, *198*, 110268. [CrossRef]
2. Al-Kadhi, N.S.; Pashameah, R.A.; Ahmed, H.A.; Alrefaee, S.H.; Alamro, F.S.; Faqih, H.H.; Mwafy, E.A.; Mostafa, A.M. Preparation of NiO/MWCNTs nanocomposite and its application for cadmium ion removal from aqueous solutions. *J. Mater. Res. Technol.* **2022**, *19*, 1961–1971. [CrossRef]
3. Mwafy, E.A.; Mostafa, A.M. Tailored MWCNTs/SnO_2 decorated cellulose nanofiber adsorbent for the removal of Cu (II) from waste water. *Radiat. Phys. Chem.* **2020**, *177*, 109172. [CrossRef]
4. Alkallas, F.H.; Toghan, A.; Ahmed, H.A.; Alrefaee, S.H.; Pashameah, R.A.; Alrebdi, T.A.; Mwafy, E.A.; Mstafa, A.M. Catalytic performance of NiO nanoparticles decorated carbon nanotubes via one-pot laser ablation method against methyl orange dye. *J. Mater. Res. Technol.* **2022**, *18*, 3336–3346. [CrossRef]
5. Jaleh, B.; Nasrollahzadeh, M.; Mohazzab, B.F.; Eslamipanah, M.; Sajjadi, M.; Ghafuri, H. State-of-the-art technology: Recent investigations on laser-mediated synthesis of nanocomposites for environmental remediation. *Ceram. Int.* **2021**, *47*, 10389–10425. [CrossRef]
6. Yu, B.; Chen, Y. Conductive WO3-x@ CNT networks for efficient Li-S batteries. *IOP Conf. Ser. Mater. Sci. Eng.* **2020**, *892*, 012027. [CrossRef]
7. Chen, J.; Wu, X.; Tan, Q.; Chen, Y. Designed synthesis of ultrafine NiO nanocrystals bonded on a three dimensional graphene framework for high-capacity lithium-ion batteries. *New J. Chem.* **2018**, *42*, 9901–9910. [CrossRef]

8. Zhang, Z.; Zhang, X.; You, X.; Zhang, M.; Walle, M.D.; Wang, J.; Li, Y.; Liu, Y.-N. 3D well-interconnected NiO–graphene–carbon nanotube nanohybrids as high-performance anode materials for Li-ion batteries. *J. Nanopart. Res.* **2016**, *18*, 247. [CrossRef]
9. Li, T.; Ni, S.; Lv, X.; Yang, X.; Duan, S. Preparation of NiO–Ni/natural graphite composite anode for lithium ion batteries. *J. Alloy Compd.* **2013**, *553*, 167–171. [CrossRef]
10. Xiao, J.; Huang, J.; Wang, M.; Huang, M.; Wang, Y. The fate and long-term toxic effects of NiO nanoparticles at environmental concentration in constructed wetland: Enzyme activity, microbial property, metabolic pathway and functional genes. *J. Hazard. Mater.* **2021**, *413*, 125295. [CrossRef]
11. Oliveira, A.R.; Correia, A.A.; Rasteiro, M.G. Heavy Metals Removal from Aqueous Solutions by Multiwall Carbon Nanotubes: Effect of MWCNTs Dispersion. *Nanomaterials* **2021**, *11*, 2082. [CrossRef] [PubMed]
12. Egbosiuba, T.C.; Abdulkareem, A.S.; Kovo, A.S.; Afolabi, E.A.; Tijani, J.O.; Bankole, M.T.; Bo, S.; Roos, W.D. Adsorption of Cr(VI), Ni(II), Fe(II) and Cd(II) ions by KIAgNPs decorated MWCNTs in a batch and fixed bed process. *Sci. Rep.* **2021**, *11*, 75. [CrossRef] [PubMed]
13. Carolin, C.F.; Kumar, P.S.; Saravanan, A.; Joshiba, G.J.; Naushad, M. Efficient techniques for the removal of toxic heavy metals from aquatic environment: A review. *J. Environ. Chem. Eng.* **2017**, *5*, 2782–2799. [CrossRef]
14. Soltani, R.; Pelalak, R.; Pishnamazi, M.; Marjani, A.; Albadarin, A.B.; Sarkar, S.M.; Shirazian, S. A novel and facile green synthesis method to prepare LDH/MOF nanocomposite for removal of Cd(II) and Pb(II). *Sci. Rep.* **2021**, *11*, 1609. [CrossRef]
15. Arsalani, N.; Bazazi, S.; Abuali, M.; Jodeyri, S. A new method for preparing ZnO/CNT nanocomposites with enhanced photocatalytic degradation of malachite green under visible light. *J. Photochem. Photobiol. A Chem.* **2020**, *389*, 112207. [CrossRef]
16. Hareesh, K.; Sunitha, D.V.; Dhole, S.D.; Bhoraskar, V.N.; Phase, D.M.; Williams, J. One-step gamma radiation aided diffusion of Ag-Au alloy nanoparticles into polycarbonate and its application towards the reduction of 4-Nitrophenol. *Radiat. Phys. Chem.* **2019**, *162*, 126–130. [CrossRef]
17. Alwan, S.H.; Alshamsi, H.A. In situ synthesis NiO/F-MWCNTs nanocomposite for adsorption of malachite green dye from polluted water. *Carbon Lett.* **2022**. [CrossRef]
18. Mwafy, E.A.; Gaafar, M.S.; Mostafa, A.M.; Marzouk, S.Y.; Mahmoud, I.S. Novel laser-assisted method for synthesis of SnO_2/MWCNTs nanocomposite for water treatment from Cu (II). *Diam. Relat. Mater.* **2021**, *113*, 108287. [CrossRef]
19. Chen, H.; Zhao, J.; Zhong, A.; Jin, Y. Removal capacity and adsorption mechanism of heat-treated palygorskite clay for methylene blue. *Chem. Eng. J.* **2011**, *174*, 143–150. [CrossRef]
20. Taherian, Z.; Shahed Gharahshiran, V.; Khataee, A.; Meshkani, F.; Orooji, Y. Comparative study of modified Ni catalysts over mesoporous CaO-Al_2O_3 support for CO_2/methane reforming. *Catal. Commun.* **2020**, *145*, 106100. [CrossRef]
21. Faghihinezhad, M.; Baghdadi, M.; Shahin, M.S.; Torabian, A. Catalytic ozonation of real textile wastewater by magnetic oxidized g-C_3N_4 modified with Al2O3 nanoparticles as a novel catalyst. *Sep. Purif. Technol.* **2022**, *283*, 120208. [CrossRef]
22. Gaudin, M.; Carles, P.; Laborde, E.; Champeaux, C.; Dumas-Bouchiat, F. A dual nanosecond-pulsed laser setup for nanocomposite synthesis—Ag nanoparticles in Al_2O_3/VO_2 matrix. *J. Appl. Phys.* **2019**, *125*, 054301. [CrossRef]
23. Fernandes, H.R.; Kapoor, S.; Patel, Y.; Ngai, K.; Levin, K.; Germanov, Y.; Krishtopa, L.; Kroeker, S.; Goel, A. Composition-structure-property relationships in Li_2O–Al_2O_3–B_2O_3 glasses. *J. Non-Cryst. Solids* **2018**, *502*, 142–151. [CrossRef]
24. Hu, H.; Saniger, J.; Castaño, V.M. Characterization of the mechanical properties of polyacrylic acid-metal oxide concretes. *Mater. Lett.* **1992**, *14*, 83–87. [CrossRef]
25. Trabelsi, A.B.G.; Alkallas, F.H.; Ziouche, A.; Boukhachem, A.; Ghamnia, M.; Elhouichet, H. Structural Defect Impact on Changing Optical Response and Raising Unpredicted Ferromagnetic Behaviour in (111) Preferentially Oriented Nanocrystalline NiO Films. *Crystals* **2022**, *12*, 692. [CrossRef]
26. Ding, S.; Fang, D.; Pang, Z.; Luo, B.; Kuang, L.; Wang, H.; Zhang, Q.; Shen, Q.; Ji, F. Immobilization of powdery calcium silicate hydrate via PVA covalent cross-linking process for phosphorus removal. *Sci. Total Environ.* **2018**, *645*, 937–945. [CrossRef]
27. Hassan, H.; Abidin, Z.H.Z.; Chowdhury, F.; Arof, A.K. A high efficiency chlorophyll sensitized solar cell with quasi solid PVA based electrolyte. *Int. J. Photoenergy* **2016**, *2016*, 3685210. [CrossRef]
28. Aslam, M.; Basit, M.; Ahmad, M.; Raza, Z.A. Structural and Band Structure Investigation of Iron Oxide Nanoparticles Incorporated PVA Nanocomposite Films. *Res. Sq.* **2021**. [CrossRef]
29. Mostafa, A.M.; Menazea, A.A. Polyvinyl Alcohol/Silver nanoparticles film prepared via pulsed laser ablation: An eco-friendly nano-catalyst for 4-nitrophenol degradation. *J. Mol. Struct.* **2020**, *1212*, 128125. [CrossRef]
30. Li, Y.; Lu, H.; Wang, Y.; Li, X. Deposition of Au nanoparticles on PDA-functionalized PVA beads as a recyclable catalyst for degradation of organic pollutants with NaBH4 in aqueous solution. *J. Alloys Compd.* **2019**, *793*, 115–126. [CrossRef]
31. ElFaham, M.M.; Okil, M.; Mostafa, A.M. Fabrication of magnesium metallic nanoparticles by liquid-assisted laser ablation. *JOSA B* **2020**, *37*, 2620–2625. [CrossRef]
32. Mostafa, A.M.; Mwafy, E.A. The effect of laser fluence for enhancing the antibacterial activity of NiO nanoparticles by pulsed laser ablation in liquid media. Environ. Nanotechnol. *Monit. Manag.* **2020**, *14*, 100382. [CrossRef]
33. Mostafa, A.M.; Mwafy, E.A. Laser-assisted for preparation Ag/CdO nanocomposite thin film: Structural and optical study. *Opt. Mater.* **2020**, *107*, 110124. [CrossRef]
34. Mostafa, A.M.; Mwafy, E.A. Synthesis of ZnO/CdO thin film for catalytic degradation of 4-nitrophenol. *J. Mol. Struct.* **2020**, *1221*, 128872. [CrossRef]

35. Lam, J.; Lombard, J.; Dujardin, C.; Ledoux, G.; Merabia, S.; Amans, D.J.A.P.L. Dynamical study of bubble expansion following laser ablation in liquids. *Appl. Phys. Lett.* **2016**, *108*, 074104. [CrossRef]
36. Khan, S.Z.; Liu, Z.; Li, L. Characteristics of γ-Al_2O_3 nanoparticles generated by continuous-wave laser ablation in liquid. *Appl. Phys. A* **2010**, *101*, 781–787. [CrossRef]
37. Piriyawong, V.; Thongpool, V.; Asanithi, P.; Limsuwan, P. Preparation and characterization of alumina nanoparticles in deionized water using laser ablation technique. *J. Nanomater.* **2012**, *2012*, 2. [CrossRef]
38. Riahi, A.; Khamlich, S.; Balghouthi, M.; Khamliche, T.; Doyle, T.B.; Dimassi, W.; Guizani, A.; Maaza, M. Study of thermal conductivity of synthesized Al_2O_3-water nanofluid by pulsed laser ablation in liquid. *J. Mol. Liq.* **2020**, *304*, 112694. [CrossRef]
39. Jafari Eskandari, M.; Shafyei, A.; Karimzadeh, F. One-step fabrication of Au@Al_2O_3 core-shell nanoparticles by continuous-wave fiber laser ablation of thin gold layer on aluminum surface: Structural and optical properties. *Opt. Laser Technol.* **2020**, *126*, 106066. [CrossRef]
40. Menazea, A.A.; Mostafa, A.M.; Al-Ashkar, E.A. Effect of nanostructured metal oxides (CdO, Al_2O_3, Cu_2O) embedded in PVA via Nd:YAG pulsed laser ablation on their optical and structural properties. *J. Mol. Struct.* **2020**, *1203*, 127374. [CrossRef]
41. ElFaham, M.M.; Okil, M.; Mostafa, A.M. Effects of post-laser irradiation on the optical and structure properties of Al_2O_3 nanoparticles produced by laser ablation. *J. Appl. Phys.* **2020**, *128*, 153104. [CrossRef]
42. Li, X.; Hu, X.; Fu, Y.; Ai, H.; Fu, M.-L.; Yuan, B. Removal of phosphate at low concentration from water by porous PVA/Al_2O_3 composites. *Environ. Technol.* **2022**, *43*, 345–354. [CrossRef] [PubMed]
43. Rahman Khan, M.M.; Akter, M.; Amin, M.K.; Younus, M.; Chakraborty, N. Synthesis, Luminescence and Thermal Properties of PVA–ZnO–Al2O3 Composite Films: Towards Fabrication of Sunlight-Induced Catalyst for Organic Dye Removal. *J. Polym. Environ.* **2018**, *26*, 3371–3381. [CrossRef]
44. Karimi, Z.; Khalili, R.; Ali Zazouli, M. Surface modified polythiophene/Al_2O_3 and polyaniline/Al_2O_3 nanocomposites using poly (vinyl alcohol) for the removal of heavy metal ions from water: Kinetics, thermodynamic and isotherm studies. *Water Sci. Technol.* **2021**, *84*, 182–199. [CrossRef] [PubMed]
45. Mukwevho, N.; Gusain, R.; Fosso-Kankeu, E.; Kumar, N.; Waanders, F.; Ray, S.S. Removal of naphthalene from simulated wastewater through adsorption-photodegradation by ZnO/Ag/GO nanocomposite. *J. Ind. Eng. Chem.* **2020**, *81*, 393–404. [CrossRef]
46. Menazea, A.A.; Ezzat, H.A.; Omara, W.; Basyouni, O.H.; Ibrahim, S.A.; Mohamed, A.A.; Tawfik, W.; Ibrahim, M.A. Chitosan/graphene oxide composite as an effective removal of Ni, Cu, As, Cd and Pb from wastewater. *Comput. Theor. Chem.* **2020**, *1189*, 112980. [CrossRef]
47. Zulfiqar, M.; Lee, S.Y.; Mafize, A.A.; Kahar, N.A.M.A.; Johari, K.; Rabat, N.E. Efficient Removal of Pb(II) from Aqueous Solutions by Using Oil Palm Bio-Waste/MWCNTs Reinforced PVA Hydrogel Composites: Kinetic, Isotherm and Thermodynamic Modeling. *Polymers* **2020**, *12*, 430. [CrossRef]
48. Inagaki, M.; Fujita, K.; Takeuchi, Y.; Oshida, K.; Iwata, H.; Konno, H. Formation of graphite crystals at 1000–1200 °C from mixtures of vinyl polymers with metal oxides. *Carbon* **2001**, *39*, 921–929. [CrossRef]
49. Shojaei, B.; Miri, R.; Bazyari, A.; Thompson, L.T. Asphaltene adsorption on MgO, CaO, SiO_2, and Al_2O_3 nanoparticles synthesized via the Pechini-type Sol−Gel method. *Fuel* **2022**, *321*, 124136. [CrossRef]
50. Hema, M.; Selvasekarapandian, S.; Arunkumar, D.; Sakunthala, A.; Nithya, H.F.T.I.R. FTIR, XRD and ac impedance spectroscopic study on PVA based polymer electrolyte doped with NH4X (X = Cl, Br, I). *J. Non-Cryst. Solids* **2009**, *355*, 84–90. [CrossRef]
51. Darwish, A.M.; Eisa, W.H.; Shabaka, A.A.; Talaat, M.H. Synthesis of nano-cadmium sulfide by pulsed laser ablation in liquid environment. *Spectrosc. Lett.* **2015**, *48*, 638–645. [CrossRef]
52. Milanović, P.; Vuksanović, M.M.; Mitrić, M.; Stojanović, D.B.; Kojović, A.; Rogan, J.R.; Jančić-Heinemann, R. Electrospun alumina fibers doped with ferric and magnesium oxides. *Sci. Sinter.* **2018**, *50*, 77–83. [CrossRef]
53. Singh, D.; Gautam, R.K.; Kumar, R.; Shukla, B.K.; Shankar, V.; Krishna, V. Citric acid coated magnetic nanoparticles: Synthesis, characterization and application in removal of Cd(II) ions from aqueous solution. *J. Water Process Eng.* **2014**, *4*, 233–241. [CrossRef]
54. Zhou, Q.; Wang, X.; Liu, J.; Zhang, L. Phosphorus removal from wastewater using nano-particulates of hydrated ferric oxide doped activated carbon fiber prepared by Sol–Gel method. *Chem. Eng. J.* **2012**, *200–202*, 619 626. [CrossRef]
55. Singh, D.K.; Mishra, S. Synthesis, characterization and removal of Cd(II) using Cd(II)-ion imprinted polymer. *J. Hazard. Mater.* **2009**, *164*, 1547–1551. [CrossRef] [PubMed]
56. Chaudhry, A.N.; Billingham, N.C. Characterisation and oxidative degradation of a room-temperature vulcanised poly(dimethyls iloxane) rubber. *Polym. Degrad. Stab.* **2001**, *73*, 505–510. [CrossRef]
57. Mwafy, E.A.; Mostafa, A.M. Efficient removal of Cu (II) by SnO_2/MWCNTs nanocomposite by pulsed laser ablation method. *Nano-Struct. Nano-Objects* **2020**, *24*, 100591. [CrossRef]
58. Xie, J.; Lin, Y.; Li, C.; Wu, D.; Kong, H. Removal and recovery of phosphate from water by activated aluminum oxide and lanthanum oxide. *Powder Technol.* **2015**, *269*, 351–357. [CrossRef]
59. Hasan, Z.; Jhung, S.H. Removal of hazardous organics from water using metal-organic frameworks (MOFs): Plausible mechanisms for selective adsorptions. *J. Hazard. Mater.* **2015**, *283*, 329–339. [CrossRef]
60. Dias, E.M.; Petit, C. Towards the use of metal–organic frameworks for water reuse: A review of the recent advances in the field of organic pollutants removal and degradation and the next steps in the field. *J. Mater. Chem. A* **2015**, *3*, 22484–22506. [CrossRef]

61. Acelas, N.Y.; Martin, B.D.; López, D.; Jefferson, B. Selective removal of phosphate from wastewater using hydrated metal oxides dispersed within anionic exchange media. *Chemosphere* **2015**, *119*, 1353–1360. [CrossRef] [PubMed]
62. Dubey, S.; Banerjee, S.; Upadhyay, S.N.; Sharma, Y.C. Application of common nano-materials for removal of selected metallic species from water and wastewaters: A critical review. *J. Mol. Liq.* **2017**, *240*, 656–677. [CrossRef]
63. Yi, R.; Guo, L.; Zou, X.; Li, J.; Hao, Z.; Yang, X.; Li, X.; Zeng, X.; Lu, Y.F. Background removal in soil analysis using laser-induced breakdown spectroscopy combined with standard addition method. *Opt. Express* **2016**, *24*, 2607–2618. [CrossRef]
64. Guo, P.; Tang, L.; Tang, J.; Zeng, G.; Huang, B.; Dong, H.; Zhang, Y.; Zhou, Y.; Deng, Y.; Ma, L. Catalytic reduction–adsorption for removal of p-nitrophenol and its conversion p-aminophenol from water by gold nanoparticles supported on oxidized mesoporous carbon. *J. Colloid Interface Sci.* **2016**, *469*, 78–85. [CrossRef]
65. Gao, Q.; Xu, J.; Bu, X.-H. Recent advances about metal–organic frameworks in the removal of pollutants from wastewater. *Coord. Chem. Rev.* **2019**, *378*, 17–31. [CrossRef]
66. Arizavi, A.; Mirbagheri, N.S.; Hosseini, Z.; Chen, P.; Sabbaghi, S. Efficient removal of naphthalene from aqueous solutions using a nanoporous kaolin/Fe_3O_4 composite. *Int. J. Environ. Sci. Technol.* **2020**, *17*, 1991–2002. [CrossRef]
67. Abbas, S.; Nasreen, S.; Haroon, A.; Ashraf, M.A. Synhesis of Silver and Copper Nanoparticles from Plants and Application as Adsorbents for Naphthalene decontamination. *Saudi J. Biol. Sci.* **2020**, *27*, 1016–1023. [CrossRef]
68. Gao, R.; Fu, Q.; Hu, H.; Wang, Q.; Liu, Y.; Zhu, J. Highly-effective removal of Pb by co-pyrolysis biochar derived from rape straw and orthophosphate. *J. Hazard. Mater.* **2019**, *371*, 191–197. [CrossRef]

MDPI AG
Grosspeteranlage 5
4052 Basel
Switzerland
Tel.: +41 61 683 77 34

Membranes Editorial Office
E-mail: membranes@mdpi.com
www.mdpi.com/journal/membranes

www.ingramcontent.com/pod-product-compliance
Lightning Source LLC
Chambersburg PA
CBHW062101220926
51850CB00010B/2852
9783725875979